THE
LITTLE BOOK
of the
BIG BANG

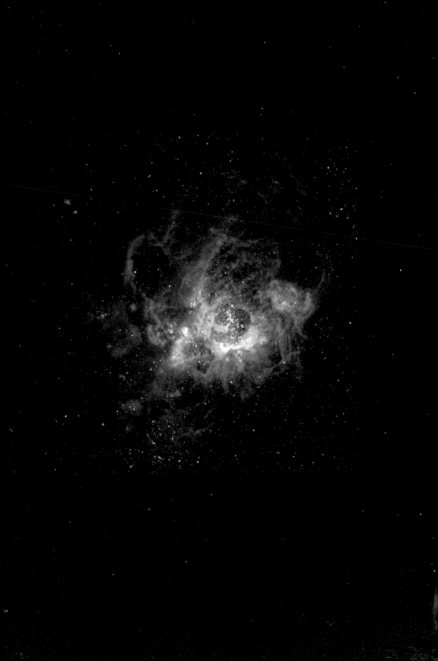

THE
LITTLE BOOK

of the

BIG BANG

A Cosmic Primer

CRAIG J. HOGAN
WITH A FOREWORD BY MARTIN REES

COPERNICUS
AN IMPRINT OF SPRINGER-VERLAG

© 1998 Springer-Verlag New York, Inc.

Published in the United States by Copernicus, an imprint of Springer-Verlag New York, Inc.

Copernicus
Springer-Verlag New York, Inc.
175 Fifth Avenue
New York, NY 10010

Library of Congress Cataloging-in-Publication Data
Hogan, Craig J.
    The little book of the big bang : a cosmic primer / Craig J.
Hogan.
        p.   cm.
    ISBN 0-387-98385-6 (hardcover : alk. paper)
    1. Cosmology.   2. Expanding universe.   I. Title.
QB981.H685   1998
523.1′8—dc21   97-39340

Manufactured in the United States of America.
Printed on acid-free paper.

9 8 7 6 5 4 3 2 1

ISBN 0-387-98385-6   SPIN 10656798

In memory of David Schramm,

cosmologist, colleague, and friend

Foreword

Speculations about our place in the cosmos are as old as thought itself. Our ancestors could weave their theories almost unencumbered by any facts—indeed that situation persisted until quite recently. But technical advances since the 1960s have transformed cosmology from speculation to serious science, vastly expanding our horizons in space and time.

Our Sun is an ordinary star among the hundred billion that make up the Milky Way—our 'home' galaxy—which is itself similar to millions of other galaxies visible with large telescopes. Just as biologists can delineate how life has evolved during the Earth's history, so astronomers are setting our entire Solar System in a cosmic context. We cannot fully understand our own origins without broadening our horizons—we are ourselves part of the cosmos. Cosmology is a 'fundamental' science, but it is also the grandest of the environmental sciences. Every atom of carbon and oxygen on Earth (and in us) was forged inside stars that died before our Solar System formed. We are stardust: or, less romantically, the nuclear waste from stars. But where did the original fuel come from? What were the embryos from which galaxies formed? The answers lies in the Big Bang, the theme of Craig Hogan's timely book.

Some of the faint smudges of light detected by powerful telescopes are actually entire galaxies so far away that the light now reaching us set out ten billion years ago. We thus have 'snapshots' showing what galaxies looked like when they were newly formed. Other delicate measurements reveal relics of still earlier eras, just a few seconds after the Big Bang itself, when everything was squeezed hotter and denser than the center of a star.

Hogan presents, from the perspective of a physicist, the modern story of cosmic evolution—how our universe has evolved from a 'Big Bang' to the intricate cosmos we see around us and of which we are part. The entire physical world—not just atoms but stars and people as well—is essentially determined by gravity and by a few basic numbers that govern microphysics: the masses of electrons and protons and the strength of the forces that bind them together and govern their motion. There are intimate connections between the cosmos and the microworld. Our everyday world is determined by chemistry, the properties of atoms. Stars shine because of reactions of atomic nuclei. Galaxies may be held together by the gravity of huge swarms of subnuclear particles.

Foreword

It is one of the triumphs of modern science—a collective achievement of thousands of astronomers, physicists and engineers, using many different techniques—that this story can now be presented with compelling conviction. Some readers, mindful of the boisterous and speculative character of earlier cosmological debates, may approach this book with scepticism, suspecting that current theories are likewise fragile and evanescent. But Hogan explains why the Big Bang is now a firm tenet of science.

Centuries ago, terrestrial maps had blurred boundaries, where the cartographers wrote 'there be dragons.' But after the pioneer navigators had traversed the globe, delineating the main land masses, there was no expectation of a new continent, nor that we'd ever drastically revise our estimates of the Earth's size and shape. In the last decade, we have, remarkably, reached this crucial stage in mapping out our universe. The occurrence of a 'Big Bang' 10–15 billion years ago is now as firmly established as (for instance) the geological inference that Europe and North America were joined together 200 million years ago. Indeed, cosmological data are now more quantitative and precise than the evidence geologists can offer.

The 1990s are the decade when the broad cosmic picture came into a clear focus: the present generation of cosmologists is fortunate. But what is perhaps even more remarkable is that cosmology has progressed at all. "The most incomprehensible thing about the universe is that it is comprehensible" is one of Einstein's best-known aphorisms, expressing his amazement that the laws of physics, which our minds are somehow attuned to understand, apply not just here on Earth but in the remotest galaxy. Newton taught us that the same force that makes apples fall holds the moon and planets in their course. We now know that this same force binds the galaxies, pulls some stars in to black holes, and may eventually cause the entire firmament to collapse on top of us. And the atoms in the most distant galaxies are identical to those we can study in our laboratories. All parts of the universe seem to be evolving in a similar way, as though they shared a common origin.

When the universe was one second old, it was almost featureless: its essence could have been described just by a few basic numbers. These numbers, plus the laws of microphysics, give the recipe for 'setting up' a universe, which could evolve, over ten billion years, from simple beginning into our present complex cosmos.

But as Craig Hogan explains, this progress brings into focus a new set of mysteries. Indeed, several basic questions still flummox us: Why is our Universe expanding? How, from its dense beginnings, did it heave itself up to such a vast size? The answer lies in exotic processes in the first tiny fraction of a second after the Big Bang, when conditions were so extreme that the relevant physics is not understood. The ultimate synthesis between the cosmos and the quantum still eludes us.

The physicist Max Planck claimed that theories are never abandoned until their proponents are all dead—that's too cynical, even in cosmology! Some debates have been settled; some earlier issues are no longer controversial; many of us have changed our minds often.

Craig Hogan has himself been an influential voice in these debates, through his physical insight and his clarity of exposition. These qualities shine through clearly in his book. He offers a balanced perspective, with a sure touch in distinguishing strongly based claims from those on the speculative fringe. He focuses on the key ideas, avoiding technical details that none but specialists need bother about. His superbly lucid account of current cosmology sets the stage for the crescendo of discoveries that seems set to continue into the next millennium.

Martin Rees,
Cambridge, England
October 1997

Contents

1 The Big Picture

We now have a model of physical events spanning the whole of observable space and time, encompassing in a complete and sweeping vision the entirety of creation since the beginning.

We live in a galaxy, in an expanding universe of galaxies, that emerged from a hot, dense early universe filled with light. All the matter of the universe was created from this light energy, and quite possibly the expanding universe itself and the energy that fills it all exploded from a tiny speck of unstable space. The nearly structure-less early universe sowed the seeds of complex structures, including galaxies and ourselves. That is a synopsis of the Big Bang, the truest model we have of cosmic evolution on the largest scales of space and time; this book tells why we think it's true.

Everyone reading this book already knows that Earth is round. This important fact about the world is not obvious, and even though it was discovered by careful thinkers in ancient times, it did not become common knowledge until a few centuries ago. The hot Big Bang model of the universe in its modern form is only about 30 years old. Increasingly precise tests have since made it one of the most firmly established paradigms in science, yet it is still widely regarded as "just a theory," and even those who are familiar with the ideas are often uncertain about the observational evidence for them. My hope is that after finishing this book, readers will know in broad terms where we are in space and time, how we got here, and how we know about it.

It's hard to shake the impression that the Big Bang is some totally bizarre thing, an extreme event, a big explosion somewhere way out in space, like something in a science fiction epic. There is even a kernel of truth in this impression. The universe does include very hot places, very dense ones, places with very strong gravity, and events very far away in space and time from where we are. But the Big Bang is mainly about our own past, present, and future—right here where we are. The compelling achievement of the Big Bang model is that it connects the familiar world to the absolute extremities of the universe in which we are embedded. It is interesting to understand the extremities but even more interesting to relate to them personally.

Chapter 1 The Big Picture

Sitting on Earth, one is not immediately aware of being part of the Big Bang. But even the fact that day is light and night is dark is evidence that the universe is neither infinitely old nor infinitely large. The sun burns because it is made of the most cosmically abundant element, hydrogen, which releases energy when converted to helium deep in its center. Stars can shine because the hot early universe created mostly hydrogen. At night, the sky is dark both because the universe is young, a condition that limits the visible volume of space and hence the number of stars available to brighten the firmament, and because the universe is expanding, so that the light that does fill the sky, light left over from the hot early Big Bang itself, has changed its color to cool, invisible microwaves. Even deep space is not perfectly cold; if your eyes could see microwaves, even at night you would see the entire sky glowing like a hot brand with leftover light from the first moments of creation. The evidence for the Big Bang is all around us if we know what to look for.

Although it is not obvious to the casual observer, the Big Bang is as real as the sun is—in a deep and precise sense, it even looks a lot like a cooler, inside-out version of the sun. But at the same time, the Big Bang model is just that: a simplified "big picture" type of idealization, like the model of Earth as a perfect sphere or the model of the sun as a perfectly round ball of hot gas. Although solar flares, sunspots, and so on are left out, the most important things about the sun—its mass, size, brightness, and color—are accurately predicted by the simplified spherical model. And for many purposes, a sphere is the best way to describe Earth;[1] the imperfections of this picture do not lead us to deny that Earth is round. Similarly, cosmological theory invokes simplifying approximations that make it possible to deal with the enormity of creation without having to attend to every detail at once. It is the combination of a simple model and precise tests that gives an approximate model the stamp of truth—the confidence that we are not just making it all up. The simplified Big Bang model works well because the universe as a whole, at early times and on large scales, is much simpler than one could guess from the complexity of the familiar world; yet at the same time the model serves as a framework supporting all the richness of natural history. It's a big canvas still filling in with interesting details.

---

[1]For many other purposes, in any situation where "up" and "down" make sense (for example, if you are trying to play tennis or soccer), the flat-Earth model is even better. As we shall see, the whole of space may similarly be globally curved but locally approximately flat.

2 A Survey of Space and Time

physics describes the contents of the universe with numbers. We usually measure the masses and sizes of everything from atoms to galaxies in some familiar unit that we can relate to, such as grams or centimeters, but all nature cares about is the ratios of scales—how many atoms are in a star, how many stars are in a galaxy, how many times an atom vibrates during the lifetime of a star. Saying that the universe is a big place is another way of saying that it contains many things. Because the universe is old, a lot of things have happened since the beginning.

How big is the universe in space and time? The universe that we can see encompasses an enormous range of scales and phenomena,[1] but it is compassed—it is not infinite. There are definite total

---

[1] The enormous range forces us to adopt "scientific notation," writing numbers in powers of ten: 10^9, for example, denotes 1,000,000,000, or a billion; 10^{-9} denotes the reciprocal of this, or one billionth.

numbers of everything: about 10^{11} galaxies, 10^{21} stars, 10^{78} atoms, 10^{88} photons. Within the observable universe that has existed since the Big Bang, there are even a definite number of distinguishable different places (about 10^{180}) and a definite number of distinguishable different times (about 10^{60}). Although the range of possible activity within this world is very large, at any given time it is not infinite.

To establish a perspective on this, it is useful to step back and survey briefly the range of things we are dealing with. There is a hierarchy of structure: Everything is composed of smaller things and is a part of something larger. The character of structures changes with scale according to the interplay of various physical forces. Quantum phenomena control the small scales, gravity dominates on large scales, and both may come into play on the largest scales. On each scale of size there is a corresponding scale of time: Things tend to happen quickly on small scales and slowly on large scales. We will find that in cosmology the largest and smallest scales of the hierarchy directly interact: Events that happened very quickly in the earliest moments of time can affect the structure and evolution of the universe on the very largest scales after a very long time.

The following surveys anticipate explanations that we will discuss throughout the book; handy summaries appear in Tables 1 and 2 (see pages 8 and 22).

Chapter 2   A Survey of Space and Time

THE MICROWORLD

In the quantum world of the very small, for each spatial size scale there is also an energy or mass carried by a particle of that size (larger energy for smaller things) and a timescale characterizing how quickly things happen. We use the conventional unit for particle energies and mass-energy (mc^2), the electron volt (about half the energy carried by a single photon of visible light). When we leave the quantum domain we measure mass by counting atoms or star masses.

The Planck scale. Energy 10^{28} eV, length 10^{-33} cm, time 10^{-43} sec. The Planck length is as many times smaller than the width of a human hair as that width is smaller than the observable universe. The Planck time is the amount of time it takes light to travel this distance. This is the smallest interval of time and space; below it, the quantum curvature effects of gravity are so large that the notion of a simple, continuous spacetime becomes inconsistent. In superstring theories that make sense of the Planck scale, this is the size of elementary particles; as one tries to delve into smaller scales, the stringy character of particles emerges, and things begin to appear larger again.[2]

---

[2] The Planck time t_p, mass m_p, and energy $m_p c^2$ are determined by fundamental constants of gravitation (Newton's constant G), quantum mechanics (Planck's constant h), and spacetime (Einstein's universal speed of light c). They reflect the scale where the quantum granularity of spacetime becomes important: $m_p = \sqrt{\hbar c / G}$, $t_p = \sqrt{\hbar G / c^5}$. Usually they are defined using the "reduced" constant, $\hbar = h/2\pi$.

Table 1. Cosmic Structures

Scale, Structure	Size	Energy, Mass	Physics and Physical Phenomena
Microworld:			
Planck scale	10^{-33} cm	10^{28} eV	smallest meaningful size; black holes merge into elementary particles
electroweak scale	10^{-15} cm	10^{11} eV	unification of weak and electromagnetic forces; origin of rest mass; lab limit
Fermi scale	10^{-13} cm	10^{8} eV	behavior of neutrons and protons in atomic nuclei; nuclear and stellar energy
Bohr scale	10^{-8} cm	10 eV	behavior of electrons bound in atoms; chemistry, periodic table
molecular world	$10^{-8} - 10^{-5}$ cm	$10 - 10^{-3}$ eV	structural combinations of atoms; proteins, DNA
Macroworld:			
everyday matter planets	10^{-8} cm–100 km	$1 - 10^{45}$ atoms	structures organized by atomic interaction
	100 km–10^{5} km	$10^{45} - 10^{54}$ atoms	atoms remain intact but overall structure shaped by gravity
degenerate bodies neutron stars	10^{5} km to 10^{4} km	$10^{54} - 10^{57}$ atoms	atoms crushed by gravity but stable; gravity crushes atoms but not nuclei
	10 km	10^{57} atoms	
black holes	size = 6 km $\times$ (mass $\div M_{sun}$)		gravity crushes all matter, liberates energy from infalling matter

Scale, Structure	Size	Energy, Mass	Physics and Physical Phenomena
Macroworld (Continued):			
normal stars	sun $= 1.4 \times 10^6$ km	$M_{sun} = 10^{57}$ atoms	gas held together by gravity, held up by heat
interstellar gas	au $= 3 \times 10^{13}$ cm pc $= 3 \times 10^{18}$ cm	$1 - 10^9$ M_{sun}	compicated intragalactic weather regulated by stellar energy
galaxies	$10^{22\pm1}$ cm	$10^{11\pm1}$ M_{sun}	self-gravitating islands of stars, gas, and dark matter
groups & clusters	10^{24-25} cm	10^{14-15} M_{sun}	galaxies in orbit around each other, clusters of dark matter
large scale structure	10^{26} cm	10^{16} M_{sun}	maximum scale of lumpiness; currently separating out of expansion
observable universe	10^{28} cm	10^{23} M_{sun}	globally uniform hypersphere; expansion controlled by gravity
beyond	$>>10^{28}$ cm	$>>10^{23}$ M_{sun}	Uniformity to very large scales, then fossil pre-inflationary structure

A summary of structures in the universe, from the quantum of spacetime to the observable universe and beyond. Each scale is characterized by a typical size or length. In the microworld of quantum mechanics, this is associated with a particle energy; in the macroscopic world we enumerate the number of atoms or the number of solar masses—bearing in mind that on the largest scales most of the mass may not be in atomic or baryonic form. Energies are quoted in units of electron volts (eV); masses in equivalent numbers of atoms or suns.

The electroweak scale. Energy 10^{11} eV, length 10^{-15} cm, time $10^{-25.5}$ sec. This is the highest energy and the shortest distance attained by human-made laboratories; it is close to the energy of interactions that give matter particles such as electrons and quarks their rest mass. The fact that it is so small means that particles need to get very close in order to interact, which makes the weak interactions that convert one type of particle to another very rare.

The Fermi scale. Energy 10^8 eV, length 10^{-13} cm, time $10^{-22.5}$ sec. This scale characterizes the realm of the strong "gluon" interactions, which determine the size and behavior as well as the masses of protons, neutrons, and atomic nuclei. The competition between gluonic and electric forces (which tend to drive the nuclei apart) creates a rich arena of nuclear phenomena and determines which stable chemical elements can exist in nature.

The Bohr scale. Energy 10 eV, size 10^{-8} cm, time $10^{-15.5}$ sec. Typical whole atoms have a size determined by the balance of electrical-attraction energy and quantum-mechanical kinetic energy. Because atoms are touching in solids and liquids, this scale determines the density and other properties of typical materials, and physics on this scale is responsible for the chemical properties of atoms. Visible light, with energies of about 2 eV, interacts strongly with atomic electrons, controlling the way solid and liquid things look.

Atomic and molecular interactions. Energies down to 10^{-3} eV, size to 10^{-5} cm or larger. Vibrations and structural transitions occur on scales from single atoms to huge collectives, such as proteins, with many thousands of atoms, capable of complex behavior.

THE MACROWORLD

Solid and liquid matter. From molecules and tiny dust particles (10^{-5} cm) to cities and small moons (about 100 km, containing about 10^{45} atoms), the shape of material bodies is determined by electrical forces between atoms and molecules. The large range of scales creates chemical and structural opportunities for extremely complex systems to develop with many levels of hierarchical organization; a person contains about 10^{28} atoms, about the number of people that would fit in a star.

Planets. In larger planets, up to about the mass and size of Jupiter (about 318 times the mass of Earth; 2×10^{30} grams, or about 10^{54} atoms; and about 140,000 km across, or 11 times the size of Earth), the atoms have enough stiffness to resist compression but are forced by gravity to adopt a roughly spherical shape.

Cold gravitating bodies. Larger cold bodies ("cold" here means that they are not stars and, like planets, are not burning nuclear

11

fuel) can maintain stable states up to about 2×10^{33} grams, or 10^{57} atoms, the mass of the sun. The larger they are, the more they exhibit the compression of gravity. Above the mass of Jupiter, gravity is so strong that it starts to compress the atoms, and each added mass causes compression to a smaller size. Extreme examples include degenerate dwarf stars with a mass about equal to that of the sun but only about as large as Earth, 10,000 km across, and neutron stars, which have a similar mass but are a thousand times smaller still.

Black holes. At still larger masses, gravity crushes all forms of matter completely, leading to collapse into black holes. Although infalling matter releases enormous amounts of energy, the gravity of a black hole traps all light emitted within its event horizon—a region about 6 km across for a hole the mass of the sun, and larger in proportion to mass for larger holes. Although black holes could in principle exist with any mass, the smallest ones probably form from stars and the largest naturally occurring holes appear to form from mass collected in the centers of galaxies; they have up to about 10^9 (a billion) times the mass of the sun, but even these are only about twenty times as large as Earth's orbit around the sun.

Normal stars. Normal stars made of hydrogen or other light nuclei avoid a catastrophic collapse for a while by nuclear burning in their cores, which releases enough heat to support them at a large size; the sun is about 1.4 million km across. These stars can be up to about a hundred times the mass of the sun.

Interstellar gas. Stars form out of larger clouds of gas and therefore tend to form together in large batches. The structures in the clouds span a wide range of scales, from the stars themselves, through the protostellar nebulae (which are about the size of the solar system; Earth's orbit, for example, is about 300 million km, or 3×10^{13} cm, across), to the much larger distances that characterize the separation between the stars. In our neighborhood, the typical distance between stars is a few light-years; astronomers like to use a unit called a parsec, 3×10^{18} cm, which is about 3 light-years.[3]

Galaxies. These "islands" of matter in the universe, which come in a wide variety of shapes and sizes, are clusters of many millions of stars orbiting each other, together with gas, in a huge swarm contained by their mutual gravity and that of some unseen material. Galaxies such as ours have about 10^{11} stars and are about 30,000 parsecs (10^{23} cm) across. Galaxies tend to orbit each other in groups and even clusters of galaxies, which may contain up to 1000 members and be up to a few million parsecs (10^{25} cm) across.

---

[3]The speed of light is 3×10^{10} cm/sec, so in one year (3×10^7 sec) it travels about 0.9×10^{18} cm; in 10 Gy, it travels about 10^{28} cm.

The expanding universe. Above a maximum scale of about 10^{26} cm, things are uniformly distributed and uniformly expanding. The entire observable universe is about 10^{28} cm across and contains about 10^{23} times the mass of the sun; on this scale, things become very smooth and structureless—much the same anywhere as anywhere else, at least as far as we can see.

THE HISTORY OF EVERYTHING

What actually happens in physics depends on what you start with. The stability and chemistry of atoms and molecules, and how fast reactions happen between them in a given situation, all follow from physics. On the other hand, the actual mix of nuclei and their spatial arrangement (into systems such as a DNA molecule or a tree or a planet or a galaxy) depend on cosmic history—on what came before. Ultimately, the stage is set by the way things unfold in the universe as a whole. Some features of the present universe (such as forms of life) are the result of a complex evolutionary path, and they reflect this complexity in their structure; other, simpler features are direct fossils of things that happened very early in cosmic history. It is useful to outline the sequence of events that make up cosmic history and to show how they are related to the observable features of the universe today.

In the following summary, "time" refers to the time it takes the universe to double its size; for the most part, it also approximates the total elapsed time or age. "Temperature" refers to the temperature of the radiation; at early times it is customary to specify this by the typical particle energy.

We can estimate from general principles a definite relation between time and temperature,[4] even at the highest temperatures. As time progresses, the universe expands and the temperature decreases into more familiar territory.

This summary reveals trends in cosmic evolution besides the obvious trend from hot to cold. There is a gradual progression from simple to complex, from nearly smooth to profoundly lumpy, from linear to nonlinear. The simple properties originate first, and complexity develops later. Related to this is the fact that the largest-scale properties are imprinted at the earliest times, and the microscopic structure emerges only very late;[5] the universe remains microscopically smooth for most of its history. The most striking lesson from the following summary is that many identifiable pieces of evidence are available from events over an enormous stretch of cosmic time.

---

[4]This relation is $(t/t_p) \approx (kT/m_p c^2)^{-2}$, where t and kT are the time and thermal particle energy, respectively, and t_p and m_p the Planck time and mass. In the early universe, most of the density is carried by radiation rather than by matter; hot radiation is denser, causing faster expansion for hotter matter.

[5]Ironically, because the universe is expanding, any given region goes the opposite way, from small size to big size, with time. The late development of structure from galaxies onward also appears to proceed from small to large scales, so there is probably a scale in the middle that is the first to depart from uniformity.

THE VACUUM ERA

The Planck epoch (time 10^{-43} sec, temperature 10^{19} GeV).[6] This epoch can be thought of as the beginning of time, because it is the boundary of the part of existence where the concept of time makes sense at all. Time does not come in smaller pieces than this, so it makes no sense to say "before" this time, and the notions of time and space are not clearly separated for intervals smaller than this. It is likely that all observable relics of this epoch were wiped out by inflation except for the emergence of spacetime itself.

The inflationary epoch (sometime after the Planck epoch and before 10^{-10} sec, temperature 100 GeV). The physical vacuum dominates the energy, developing a repulsive gravity that drives the universe to enormous size. This could be regarded as the "Bang" it-

---

[6]In the first part of this list, I use energy units that reflect the energy of thermal motion of a typical particle; multiply the temperature in GeV by 10^{13} if you prefer to use Kelvin units (kelvins or simply K for short) for temperature. One GeV is the energy contained in the mass of a proton, 10^9 electron volts; an MeV is a million electron volts, a keV a thousand. The Kelvin scale is the same as the Celsius (centigrade) scale except that it is measured from absolute zero instead of from the freezing point of water (which lies at 273.15 K). Thus, in the natural Kelvin scale, there are no negative temperatures. Temperature has a direct physical meaning: Hotter bodies are made of faster-moving particles, with speeds in proportion to the square root of the temperature, and absolute zero corresponds to no thermal motion at all.

self inasmuch as we are still seeing its effects in the form of the cosmic expansion. It is possible, though it has not yet been proved, that this process is also responsible for the fluctuations that led to the formation of galaxies and all other inhomogeneous structures in the universe today. If so, we might find detailed relics of this time in the microwave background radiation fluctuations or primordial gravitational waves.

THE RADIATION ERA

The creation of light (after inflation and before the creation of matter, temperature above 100 GeV). The vacuum energy transforms itself and flows into the form of more familiar particles such as photons or light quanta, as well as particles and antiparticles of matter in equal numbers. This epoch is sometimes called "reheating," the conversion of energy into thermal radiation. The repulsive gravity disappears, so gravity adopts its familiar attractive behavior. The background radiation energy we see filling the universe today originates here. It is possible that cosmic dark matter is also produced as early as this time and dominates the mass of the universe today.

The creation of baryonic matter (after the creation of light and before or at the electroweak transition, temperature 100 GeV). A small excess of quarks and electrons over antiquarks and antielectrons is generated in a process called baryogenesis. This process leaves its imprint in the presence (and measurable abundance) of baryonic matter today, which includes all normal matter—all the atoms in all the stars and galaxies.

The electroweak epoch (time 10^{-10} sec, temperature 100 GeV). This watershed represents the threshold of currently laboratory-tested physics; in other words, the previous phenomena are analyzed mostly as cosmological constraints on physical theories, rather than vice versa. This epoch is significant in other ways as well. For example, it marks the shift of the vacuum to the state that gives matter particles their rest masses, it may be the epoch of baryogenesis, and it may be the end of the supersymmetry of the vacuum, the time when matter and forces become distinguishable forms of energy with different behaviors. There may be cosmic relics of this period (such as dark matter or cosmic defects), or there may not.

The strong epoch (time 10^{-4} sec, temperature 0.2 GeV). At about this time the universe makes a transition from "quark soup" to hadronic matter, where quarks and gluons adopt their familiar modern existence, hiding exclusively inside neutrons and protons. This transition may have left relics in the universe today, such as various forms of dark matter (axions, black holes, or quark nuggets), or it may have left the matter in a lumpy state that would affect the creation of nuclei later on.

Decoupling of the weak interactions (time 1 second, temperature 1 MeV). The reactions that turn neutrons and protons back and forth into each other become ineffective at this point. The most important relic of this is the fact that protons outnumber neutrons by about seven to one, which is why the universe is made mostly of hydrogen today. The numerous relic cosmic background neutrinos that have been streaming to us since this time without interaction have their density fixed at this time. Several possible forms of dark matter decouple at this time and have their abundances fixed at about this time. Shortly afterward, the last big antimatter annihilation occurs: Electrons and positrons dump their heat into the cosmic background photons, which thereby still have a bit more energy today than the neutrinos do.

Creation of nuclei of the light elements (time 100 sec, temperature 0.1 MeV). Things have cooled enough for the neutrons and protons to stick to each other in a process called nucleosynthesis. Nearly all the neutrons join with protons into helium nuclei, and nearly all the rest into deuterium nuclei, with a tiny bit remaining as lithium and practically nothing heavier. The abundances of these elements today provide a precise test of our understanding of this epoch.

Decoupling of the radiation spectrum (starting at time 1 month, temperature 500 eV). As the Universe thins out, the interaction of matter and radiation becomes less efficient. The very precise Planck spectrum of the radiation we see today is established by this time, and little extra energy could have been added afterward without being detected. The decoupling happens in stages; the process of new photon production ends early, whereas the energy exchange between particles continues a while longer. The final conversion of primordial energy into most of the cosmic background photons we see today takes place at this time.

THE MATTER ERA

The transition from the domination of radiation to that of matter (time about 10,000 years, temperature 30,000 K, or 3 eV). The matter of the universe loses less energy than radiation does to the expansion, so eventually it dominates the mass density around this time. The baryonic matter is still tightly controlled by the stiff, high pressure of hot radiation, but most of the gravity and most of the mass come from the dark matter, which can move freely starting about now. This epoch imprints the spectrum of fluctuations in the density of cosmic mass, which imprints the present-day large-scale distribution of matter.

Last scattering (time about 500,000 years, temperature about 3000 K). When the radiation cools enough for electrons to attach to protons (hydrogen "recombination"), the electrons almost stop interacting with the radiation. The radiation propagates freely through space after this. We see directly to this epoch in our cosmic background maps, so as better maps are made, many details of this epoch will be clarified. The baryonic matter, decoupled from the radiation, is free to move about after this; the suppression of cosmic structure growth by radiation ends.

The dark ages (until a time of about 1 billion years, temperature of about 0.002 eV, or 20 K). The gravity of matter acts to amplify primordial fluctuations in the dark matter and the baryons, but these structures are for a long time small-amplitude ripples. At some time during this period, the strongest ripples first break into whitecaps, forming the first nonlinear structures—the first structures to stop expanding—probably subgalactic units of perhaps a million times the mass of the sun.

The epoch of galaxy formation (starting at a time of about 1 billion years, temperature of about 0.002 eV, or 20 K). Gravity causes the ripples to grow in scale, and bigger waves break, leading to formation of the first galaxies. Gas collapses into galaxies and rapidly forms stars and quasars bright enough to see today. The light and heat from these energy sources ionizes the remaining gas.

Table 2. The History of Everything

Epoch	Time	Temperature	Physics and Physical Events	Relics and Observables
Planck epoch	10^{-43} sec	10^{19} GeV	limit of spacetime: quantum gravity, superstrings	four-dimensional spacetime
Cosmic inflation			unstable vacuum blows apart	size and shape of observable universe
			quantum fluctuations in vacuum	large-scale structural fluctuations
Creation of light			conversion of vacuum to radiation energy	energy of cosmic background radiation
Creation of matter			net baryon number generated	excess of matter over antimatter
Electroweak epoch	10^{-10} sec	100 GeV	electroweak unification, origin of mass	differentiated force and matter fields
Strong epoch	10^{-4} sec	200 MeV	quark $\rightarrow$ hadron plasma, neutrons and protons form	exotic forms of dark matter
Weak decoupling	1 sec	1 MeV	neutrinos decouple, neutron/proton ratio fixed	universe dominated by hydrogen nuclei
$e^-\,e^+$ Annihilation	5 sec	0.5 MeV	electron heat dumped into photons	photons hotter than neutrinos today
Nucleosynthesis	100 sec	100 keV	nuclear reactions freeze out, nuclei form	light element abundances: He, D, Li

The History of Everything, *continued*

Epoch	Time	Temperature	Physics and Physical Events	Relics and Observables
Spectral decoupling	1 month	500 eV	end of efficient photon production	blackbody spectrum of background radiation
Matter/radiation equality	10,000 years	30,000 K $\approx$ 3eV	matter dominates mass density	character of fluctuations in mass density
Matter/radiation decoupling	0.5 My	3,000 K	hydrogen recombines, Universe transparent to light	fluctuations in background radiation
Dark Ages	$\lesssim$ 1 Gy	$\gtrsim$ 20 K	mass fluctuations grow, first small objects coalesce	first stars, heavy elements
Galaxy formation	$\gtrsim$ 1-2 Gy	$\lesssim$ 10-20 K	collapse of galactic systems, quasar engines	stars, galaxies, quasars directly visible
Bright Ages	2-13 Gy	3-10 K	gas consumed into stars, remnants, planets	formation of Milky Way and Solar System
Present epoch	13 Gy	3 K	large scale gravitational instability	superclustering of galaxies
Future	$>>10^{11}$y	$<<1$ K	unpredictable runaway complexity	up to us

A summary of important epochs in the history of the universe from the first moments of time. Each epoch is characterized by the time it takes the universe to double its size, which approximates the age. The temperature steadily decreases as the universe expands. Gy = gigayear, or 1 billion (10^9) years; eV = electron volt, a unit of particle energy; K = Kelvin temperature scale.

The bright ages (until the present time, about 13 billion years, temperature 2.7 K). The primordial gas is gradually consumed and increasingly polluted by heavy elements ejected from stars. Over the last few billion years, the flurry of star formation has been fading; quasars have almost disappeared, and the formation of new galaxies has become quite rare. During the same period of time, life has evolved on our own planet and very recently has sprouted intelligence.

The future. The universe will cool, the stars will die out, and cosmic activity will slow—but the structural opportunities are open-ended and will become increasingly complex as a wider range of spacetime scales and an expanding supply of gravitational energy become available.

3 A Summary of Physics

Everything that happens in the universe consists of the same basic stuff, "mass-energy," transfigured in space and time from one form into another. Not just anything can happen to this stuff; its forms and transformations are described by the mathematical metaphors of physics that underlie the ideas of modern cosmology.

TIME AND SPACE

Everything that happens occurs in space and time. The two together, called spacetime, constitute a kind of stage upon which all the forms of energy dance. We are familiar with certain aspects of space and time from our direct experience of the world—we know, for example, that space has three large dimensions and that time is very different from space. Other important properties of spacetime are revealed only by detailed, abstract mathematical reasoning—for example, that curvature of spacetime creates gravitational forces. The local properties of spacetime determine a lot of what is possible and what is not (it is not possible, for example, to overtake a light beam). Physics generally regards space and time as continuous; they can be subdivided into infinitely small intervals, and each position or instant follows the last with no gaps. This view is consistent within limits but breaks down on extremely small scales, where space and time may be very strange indeed, perhaps foam-like or even discrete. Furthermore, the spacetime that spans the entire universe may have surprising and counterintuitive global properties that cannot be discovered by any local experiment. For example, time itself may have abrupt ends and beginnings ("space-time singularities," such as the centers of black holes), and space may fold back on itself so that it is not infinite but also has no edges ("closed" universes). Figure 1 is an example of a representation of a small piece of spacetime in diagram form.

PARTICLES AND FIELDS IN SPACETIME

Spacetime is filled with a variety of fields that carry energy through spacetime. Each of these fields—such as the electromag-

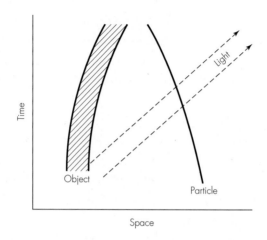

FIGURE 1 A spacetime diagram. Real spacetime is four-dimensional (the three familiar dimensions of space, plus time), but here we draw only one dimension of space, plus time. Every point is an "event," an instant of time at a particular place. A particle has a "world line" showing its position at each time, and an extended object sweeps out a "world swath." Because nothing can travel faster than light, these lines and swaths are always more nearly vertical than lines traced by light rays. Spacetime is both an arena for physics and a participant: In Einstein's theory of gravity, spacetime is curved, and this makes world lines bend, which is another way of saying that particles fall where there is gravity.

netic field or the electron field—is present everywhere, at every location in space and time, even in so-called empty space, or vacuum (the vacuum is just the lowest-energy state of the system of fields, all of which carry energy). The behavior of the fields' activity can be described as waves, vibrational energy like that carried by waves in water. The fields can be thought of as an intricate web of springs filling space, with many different types of possible oscillations. Like water waves, the oscillations of the fields can be big or small, long or short, fast or slow.

The dancing of the fields can also be described as particles, discrete packets of energy that travel about and interact with each other, bouncing off each other, changing direction, being created and destroyed, and turning into different types of particles. The terms *particles* and *fields* thus refer to different aspects of the same entities, and the two descriptions of the world are interchangeable—they express the same thing in distinct but mathematically equivalent languages.

All energy is carried by some field. One of the great discoveries of physics is that everything that happens in the universe can be explained in terms of the action of a few different types of particles (or fields), each with distinctive properties. The entire range of possible events is determined by the patterns of behavior of these fundamental natural fields. Close study has uncovered precise mathematical rules that govern this behavior.

Nature contains two classes of fields: matter and forces. Matter interacts with other matter through the intermediate action of forces. In order to connect to a force, the matter must carry the appropriate charge that couples to that force. When a force connects matter particles that are separated in time as well as in space, it is given the name *radiation;* it carries energy through space and time. Each force is associated with its own charges and its own type of radiation.

The properties of matter and forces resemble the rules of chess in that a huge variety of games can grow out of just a few simple rules. In physics, the rules include what each force can do and what charges and masses each matter particle carries, the pieces are the fields themselves, and the game is the natural world playing out in spacetime. Certain regularities appear as conservation principles or symmetries; for example, the shape of the board, the total number of pieces, and the total number of squares always stay the same no matter what happens, just like the total amount of energy.

The properties of the matter and force fields need to be supplemented with some rules of play—in addition to the moves allowed each piece, we need to know when the moves happen and what results from them. Nature follows rules called mechanics, or laws of motion. On large scales, the rules of classical Newtonian mechanics state, for example, that if a force acts on a particle, then that particle accelerates as a result, and that if no force acts on it, then its motion remains unchanged. On very small scales, the rules of quantum mechanics dictate additionally that rapid time changes or rapid space changes always correspond to large energy. This means that, for example, a particle confined to a very small space must move very quickly, so a large force is required to keep it from getting away. Quantum mechanics connects the particle and wave descriptions of things: Quickly moving particles have more energy and are associated with short waves and rapid oscillations, and slowly moving or lower-energy particles are associated with long waves and slow oscillations. The structure of spacetime also guarantees that if a certain type of particle exists, then so does another type of particle, of equal mass but with all the opposite charges, called an antiparticle. A particle and its antiparticle can meet and annihilate, converting all of their mass into energy carried away by radiation and leaving no matter at all.[1] (See Figure 2.)

THE FORCES OF NATURE

Only four kinds of force fields have so far been discovered: gravity, electromagnetism, the strong force, and the weak force. Each has its corresponding particles: gravitons, photons, gluons, and "W and Z

---

[1]Energy and mass are equivalent; there is no mass without energy. There is, however, another use of the term *mass* that refers to the rest mass of a particle: the energy or mass it has when it is standing still. Some particles, such as photons, can never stand still and so are said to be "massless" even though they have energy. The mass m and energy E of a particle are related by Einstein's famous formula $E = mc^2$, where c is the speed of light. An annihilation between a particle and an antiparticle liberates energy in the amount $2mc^2$. Conversely, if this much energy is available in a small space, then a particle and an antiparticle can be created out of energy alone. This is rare today, but it happened a lot in the early universe because the temperature was so high that the particle collision energies often exceeded the masses of elementary particles.

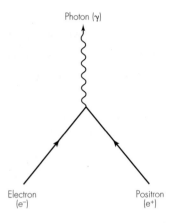

FIGURE 2 A "Feynman diagram," a way of representing quantum processes. This example shows an electron and its antiparticle, a positron, annihilating to produce a photon. The positive and negative electrical charges exactly cancel, leaving zero charge for the photon to carry off. The diagram can be thought of as representing either particles or waves.

vector bosons," respectively. Each type of force also corresponds to a type of radiation, although only two of them, gravitational radiation and electromagnetic radiation (or light), propagate over distances much larger than an atomic nucleus. In each case *particle, radiation,* and *force* are names given in different contexts to the same physical entity.

Gravity is the simplest force. To couple to gravity requires only the presence of energy; thus gravity couples all forms of energy (that is, all the fields, both forces and matter) to each other. Every particle of any kind, because it must carry energy to exist, experiences a gravitational force. For normal forms of matter, this force is always attractive: Gravity tends to make things move toward each other. (No repulsive gravity has ever been seen, although it is physically possible in some circumstances and is likely to have been critically important in starting the Big Bang.) Finally, it is long-range: no matter how far apart things are, gravity couples them.

For these reasons, gravity is the most important force between very large things—between the sun and its planets, between entire galaxies. For this reason, it also controls the expansion of the universe. Gravity was the first force with a precise mathematical formulation: Newton's law of universal gravitation showed that the same force governs apples falling to the ground, the continuous falling of the moon, and the falling of the planets in their orbits. The radiation form of gravity, gravitational waves, has not been detected directly, although its energy losses have been indirectly measured.[2]

Electromagnetism is somewhat more complicated than gravity. To couple to it requires electrical charge; unlike energy,

---

[2]Gravitational radiation is as hard to generate as to detect, but we know it exists: High-precision measurement of pulses from orbiting neutron stars reveal energy losses by gravitational radiation. Gravitational waves interact so weakly with matter that they propagate freely through the densest matter, even through the densest plasma of the early universe. Because gravitational waves are so difficult to detect, there is little hope of detecting single graviton particles.

which all particles have, electrical charge is a property that some particles have and others don't. Thus some kinds of matter, which are electrically neutral, do not experience electrical forces. Also, electrical charge comes with either of two signs, positive or negative, which can cancel each other if both are present in equal amounts. Electrical forces between particles can be either attractive (for opposite charges) or repulsive (for like charges). Magnetic forces, which are in essence the same thing as electrical forces, arise if there are *moving* charges, such as an organized electrical current inside a coil of wire or inside the aligned atoms of a compass needle. Like gravity, electromagnetic forces operate over large distances. Magnetic fields affect the solar wind and cosmic rays (and Earth's climate). They extend across our galaxy, control the formation of stars, and catalyze the extraction of energy from black holes.

All light, including radiation from radio waves through visible light to X rays, is made of jiggling electromagnetic fields—an amazing fact precisely described in the nineteenth century. All the apparent differences between these radically different types of radiation arise only from how fast or slow the jiggling is—the frequency or wavelength of the light. Note that light itself is not electrically charged; it only couples to charge. Light is made of the same electromagnetic fields that make a compass needle turn or your hair stand on end. These fields are caused by the presence and movement of electrically charged matter.

Together with the laws of quantum mechanics, electromagnetism determines the structure of the electron clouds that make up most of the volume of the atoms of all ordinary matter. It therefore controls the chemical behavior of all atoms and molecules, including the ones we are made of.

The strong force, also called the chromodynamic force, is mathematically the most complex. Although it is has important consequences for the structure of matter, it is not directly perceived in everyday life. In contrast to the single electrical charge, the strong force couples to a complex "strong charge" called color, which comes in three kinds: R ("red"), G ("green"), and B ("blue"). These charges can be mixed together in a single particle, and the force depends on the mix. Strong forces can be either attractive or repulsive; they can even be sideways—in some direction other than together or apart. Whereas there is only one kind of photon (that is, one kind of electromagnetic radiation, the different colors differing only in frequency), eight distinct types of gluons make up the strong force, eight different varieties of "strong light."

Unlike photons, gluons themselves carry the color charge that they couple to. Because photons are electrically neutral, two beams of light simply pass through each other; light never emits light, only matter does. But gluons scatter off of other gluons and emit other gluons all the time. This self-interaction makes the gluon forces effectively very short-range. They never get very far, and over a long distance, the strong forces always arrange to cancel each other out. The strong force therefore operates only at short distances and at high particle energy; it controls the sizes and shapes of the nuclei of atoms, the tiny aggregations of neutrons and protons ("nucleons") at the centers of atoms. Atomic nuclei contain most of the atoms' mass. Indeed, most of the mass is made of gluonic energy, which implies that most of the mass of familiar things is made of energy from intrinsically massless gluons whose true massless character is trapped, bouncing around inside the nucleus. (It is ironic that gluons, which are forces and not matter, make up most of the mass of familiar matter. How flexibly our thinking must adapt to unfamiliar realms of nature!) Strong forces determine which atomic nuclei in nature are stable and, therefore, which chemical elements can exist.

The weak force couples to another complex and esoteric charge called weak isospin (it is named after a mathematically related spin behavior). Unlike gravitons, photons, and gluons, the W and Z vector bosons that carry the weak force have mass, which gives this force a very short range indeed—so short that weak interactions happen extremely rarely, and "radiation" made of W and Z particles does not even travel far enough to cross an atomic nucleus. Weak forces, though subtle and slow, are critically important because they transform particles between different nearly stable forms (for example, from neutrons to protons). Such decay changes the composition of a nucleus, turning one element into another; the radioactive tritium that powers the glowing hands of a watch is decaying from a heavy form of hydrogen into a slightly lighter isotope of helium.

The similarities in character between these forces probably reflect a deeper underlying commonality. At a deeper level of understanding, all the forces may eventually be understood as different manifestations of a single force. Indeed, such a connection has already been demonstrated between the electromagnetic force and the weak force; both are manifestations of an underlying "electroweak" force. The fundamental symmetry between them is invisible now because of a phenomenon called spontaneous symmetry breaking, which forces the system of fields to collapse into a lowest-energy state that does not respect the symmetry between the forces.[3] In this state, some of the electroweak fields look like photons and others look like W and Z massive vector bosons. (An interesting by-product of this symmetry breaking is to give (rest) mass to particles that would otherwise be massless, such as the electron.) There are many ideas for a grander unification of the forces. The Grand Unified Theories, for instance, aim to unify the strong force with the electroweak force by using the same general scheme as electroweak unification. Unification with gravity seems to be much harder and will probably require a unification of forces with

---

[3]A concrete illustration is a pencil balanced on its tip on a flat table. It falls down in a particular direction. The pencil, lying on its side, now points in a particular direction even though neither the pencil nor the table has any directional asymmetry. More detail on this metaphor follows in Chapter 8.

matter—a Theory of Everything. This seems appropriate for a force that couples to all forms of energy.

TYPES OF MATTER

There are three known "families" of matter fields (also known as generations), each with the same pattern. The first family contains all of the different types (or "flavors") of particles that make up everyday matter. This family contains four types of particles: two flavors of quarks ("up" and "down"), the electron, and the electron neutrino. All atoms of all familiar matter are made of a nucleus surrounded by electrons, bound together with electromagnetic forces. Nuclei of atoms in turn are made of positive electrically charged protons and electrically neutral neutrons, which in turn are made of three quarks bound together by "gluonic" strong interactions.

Matter particles are distinguished by their masses and their charges, the forces they couple to. All of them experience gravity and weak interactions—that is, they all carry energy and weak isospin. For the electron neutrino, these are the only forces, which is why the neutrino plays an important role only in very small-scale and very large-scale phenomena: It notices short-range forces from weak interactions and long-range forces from gravity but plays no role in the vast array of cosmic electromagnetic phenomena from atoms to quasar jets. Electrons and quarks experience in addition the electromagnetic force—that is, they carry electrical charge that binds them together in atoms.

Only the quarks, however, experience strong forces—that is, quarks carry color charge, each quark flavor being available in multiple colors. Because of the special self-interacting character of the strong force, quarks are always grouped together in threes inside neutrons and protons or as pairs in pions. No free quark has ever been detected lying around loose on its own. The total number of quarks (minus the number of antiquarks) in any nucleon cannot be easily changed by any known force, which is why matter is stable and does not decay, even though it has lots of stored energy.[4] Normal matter is called baryonic matter (from a Greek root meaning "heavy"), because its mass is dominated by baryons such as neutrons and protons, which, as we have seen, are themselves made of quarks and gluons. (The lighter nonquark particles of normal matter, electrons and neutrinos, are called leptons after the Greek word for "light.") All atoms of matter in everything you see around you

---

[4]Einstein's formula $E = mc^2$, which describes the equivalence of mass and energy, shows the enormous amount of energy associated with even modest masses. An engine that extracted the whole rest energy $E = mc^2$ from its gasoline fuel would need a billion times less fuel than one that just burns that fuel normally. Even nuclear energy has an efficiency of less than 1 percent of mc^2. There are, however, systems in the universe, such as quasars and gamma ray bursters, that extract 10 percent or more of mc^2 from the gravitational energy of material falling into black holes. A car based on this technology would get a billion miles to the gallon!

are baryonic, as is the material in stars and in the atomic gas between them.

The same pattern of charges applies to the particles of the other two families of matter particles. The second family includes the "charmed" and "strange" quark flavors and the muon and its neutrino, the third family the "top" and "bottom" quarks and the tau lepton and its neutrino. Thus altogether there are six quark flavors, three electrically charged leptons, and three flavors of neutrino. The second and third families are both heavier than the first, and they decay very quickly to the first family, which is why they are not familiar in everyday life or common in the cosmos at large. They are created temporarily in particle accelerators or naturally by cosmic rays, but they have not been abundant in nature since the first microsecond of the Big Bang.

THE STANDARD MODEL AND UNIFICATION

As far as we know, this is a complete inventory of all the possible types of matter and all the possible forces between them. This litany of rules seems arbitrary, especially when stripped of its mathematical precision, but that is how nature works. The complete set of rules is called by physicists the Standard Model. Perhaps it is given this unprepossesing name in the hope that it will someday be replaced by something prettier or simpler.

Table 3 summarizes the Standard Model particles and forces. The typical phenomena listed at the bottom of the table are examples of circumstances where each force plays a dominant role. It is important to remember, however, that the Standard Model as a whole *fully describes the basic particles and interactions involved in every phenomenon that has ever been observed in the laboratory.*

It is possible to delve deeper and ask why just this particular selection of fields and rules occurs—to ask for a deeper rationale. Perhaps all of these conceptual structures (matter and forces, mechanics, space and time) can be viewed mathematically as manifestations of a single underlying structure. Mathematical theorists are working on finding such a Theory of Everything and even have a promising candidate theory, Supersymmetric Superstrings, in which everything, even the spacetime everywhere—including in particular the energetic spacetime around black holes—is described as quantum excitations of string-like objects. To be successful, any such theory must agree with the physics we know about, the Standard Model. But if past experience is any guide, the richer

---

[5]This very small number, 10^{-38}, the square of the ratio of the proton mass m_{proton} to the Planck mass m_p, leads to all the very small and very large numbers in astrophysics. For example, a star contains about $(m_p/m_{proton})^3 \simeq 10^{57}$ atoms, and lasts for about the same number of Planck times; the universe contains about $(m_p/m_{proton})^4$ atoms. The origin of this number in fundamental physics (i.e., the reason why the gravitational force is so weak) remains a mystery.

Table 3. Forces of Nature, Types of Matter and their Couplings in the Standard Model.

Force:	Gravity	Electro-magnetism	Strong Force	Weak Force
Force Particles:	gravitons	photons	8 gluons	$W^{\pm}$ and Z^0 bosons
Radiation:	gravitational waves	light	confined	short range
Coupled charge:	all energy	electrical charge	SU(3) color (R,G,B)	SU(2) weak isospin
Matter Particles:				
1 { up quark	$10^{-38} \times \left(\dfrac{\text{energy}}{m_{\text{proton}}c^2}\right)^2$	$\dfrac{2}{3} \times \dfrac{1}{137}$	10^{-1}	10^{-2}
down quark	•	$-\dfrac{1}{3} \times \dfrac{1}{137}$	•	•
electron	•	$-1 \times \dfrac{1}{137}$		•
electron neutrino	•			•
2 { charmed quark	•	•	•	•
strange quark	•	•	•	•
muon	•	•		•
muon neutrino	•			•
3 { top quark	•	•	•	•
bottom quark	•	•	•	•
tau lepton	•	•		•
tau neutrino	•			•
Phenomena:	stars, solar system, galaxies, universe	visible light, atoms, molecules, chemistry, biology	atomic nuclei, nuclear energy	radioactive decay, element formation

This chart shows all known types of matter and forces of nature. Each field and coupling corresponds to a mathematical object in a system of equations that together describe the Standard Model of fundamental interactions. The top rows indicate the fundamental forces, the corresponding particle names, the long-range radiation for gravity and electromagnetism, and the charges they couple to. The matter particles are listed at the left, and the approximate strengths of couplings to each force are indicated by the numerical entries in the table. The precise couplings depend somewhat on energy, so these entries are approximate. Note the relative weakness of the gravitational force.[5] Bullet marks (•) denote continuation of a similar pattern. Note the three similar "generations" or "families" of matter particles, only the first and lightest of which is common in ordinary matter. Each quark flavor listed here comes in all three colors. The properties of these particles and their interactions determine the properties of nuclei and atoms, shown in the periodic table. The phenomena listed at the bottom are examples of natural situations wherein each force plays an especially prominent role.

mathematical structure of the new theory will also predict new things not yet discovered.

It is therefore possible that other particles and forces will be found that supplement the Standard Model. This is another important reason to study cosmology carefully: It provides a promising arena for finding the signs of new physics. For example, new types of fields are apparently required to explain the starting conditions of the Big Bang, the fluctuations that form the seeds of large cosmic structures, and possible new types of invisible nonbaryonic matter that may comprise most of the total mass of the universe.

THE MACROSCOPIC WORLD

The particles of nature comport themselves according to the rules of the Standard Model. The possible arrangements and typical behaviors exhibited by everything from stars to quasars to hurricanes to roses are consequences of the forces of the Standard Model and their couplings to matter. All the things surrounding us in our everyday lives—including the air we breathe and our own bodies and brains—are made of atoms. The whole variety of different chemical substances arises from the variety of ways in which fundamental particles can bind into stable structures, including very complex ones such as proteins and very interesting ones such as DNA. The nature of physical interactions dictates what kinds of things can or can't happen; not everything is possible. For example, the fact that gravity is so much weaker than the other forces is responsible for the huge sizes of stars and planets. Indeed, the weakness of gravity is responsible for the very existence of a macroscopic world—why the universe can encompass a large range of scales in space and time and why it can contain more than a single atom.

The variety of stable forms arises from a dynamic balance: Some forces pull things together, but other forces, and quantum mechanics itself, tend to drive them apart. Recall that the smaller the volume confining a particle, the faster it moves, and the more force is needed to confine it. Particles explore many configurations but, if left alone, tend to end up in the stable lowest-energy ones, where the confining force is just exactly in balance with the tendency of quantum mechanics to make things fly apart. Thus attractive forces automatically form stable structures on small scales, of which nuclei and atoms are the ubiquitous examples. The electrical force holds the electron to the nucleus, but if they get too close together, the laws of quantum mechanics require them to have more energy, which resists further compression. The existence of a stable point where these tendencies balance is the reason why atoms have a size and solid objects are solid. When you come home at night, it is the laws of quantum mechanics that require you to open your door rather than simply pass through it. The mathematical regularities of the fundamental forces lead to

KEY

Atomic number
Atomic Symbol
Atomic weight*

Periodic Table of the Elements

Period	Group I	Group II											Group III	Group IV	Group V	Group VI	Group VII	Group VIII
1	1 H 1.008																	2 He 4.003
2	3 Li 6.941	4 Be 9.012											5 B 10.81	6 C 12.01	7 N 14.01	8 O 16.00	9 F 19.00	10 Ne 20.18
3	11 Na 22.99	12 Mg 24.31											13 Al 26.98	14 Si 28.09	15 P 30.97	16 S 32.06	17 Cl 35.45	18 Ar 39.95
4	19 K 39.10	20 Ca 40.08	21 Sc 44.96	22 Ti 47.90	23 V 50.94	24 Cr 52.00	25 Mn 54.94	26 Fe 55.85	27 Co 58.93	28 Ni 58.71	29 Cu 63.55	30 Zn 65.38	31 Ga 69.72	32 Ge 72.59	33 As 74.92	34 Se 78.96	35 Br 79.90	36 Kr 83.80
5	37 Rb 85.47	38 Sr 87.62	39 Y 88.91	40 Zr 91.22	41 Nb 91.22	42 Mo 95.94	43 Tc (99)	44 Ru 101.1	45 Rh 102.9	46 Pd 106.4	47 Ag 107.9	48 Cd 112.4	49 In 114.8	50 Sn 118.7	51 Sb 121.8	52 Te 127.6	53 I 126.9	54 Xe 131.3
6	55 Cs 132.9	56 Ba 137.3	57 La 138.9	72 Hf 178.5	73 Ta 180.9	74 W 183.9	75 Re 186.2	76 Os 190.2	77 Ir 192.2	78 Pt 195.1	79 Au 197.0	80 Hg 200.6	81 Tl 204.4	82 Pb 207.2	83 Bi 209.0	84 Po (210)	85 At (210)	86 Rn (222)
7	87 Fr (223)	88 Ra 226.0	89 Ac (227)															

Lanthanide series

58 Ce 140.1	59 Pr 140.9	60 Nd 144.2	61 Pm (145)	62 Sm 150.4	63 Eu 152.0	64 Gd 157.3	65 Tb 158.9	66 Dy 162.5	67 Ho 164.9	68 Er 167.3	69 Tm 168.9	70 Yb 173.0	71 Lu 175.0

Actinide series

90 Th 232.0	91 Pa 231.0	92 U 238.0	93 Np 237.0	94 Pu (244)	95 Am (243)	96 Cm (247)	97 Bk (247)	98 Cf (251)	99 Es (254)	100 Fm (253)	101 Md (256)	102 No (254)	103 Lw (257)

*The number in ( ) = Mass Number of the most stable isotope.

regularities of the stable structures they form. These regularities are reflected in the periodic table of the elements (Figure 3), which classifies the arrangements of the outermost shells of electrons in atoms.

A much larger chart could show the possible arrangements and combinations of atoms into molecules, but as the size of the molecules grows, the number of possibilities becomes enormous. This wealth of potentiality is the wellspring of life. Life is not another "fundamental force" of nature but rather arises from the incredible variety of ways in which certain atoms can interact with each other. Electron cloud patterns in the atoms of living matter form myriads of flexible but stable and versatile attachments such as hinges and loaded springs. These atoms—mostly simple protons or (electrical) charge-1 nuclei (hydrogen), charge-6 nuclei (carbon), and charge-8 nuclei (oxygen), plus smaller proportions of many other elements—thereby construct the elaborate molecular

---

FIGURE 3 The periodic table of the elements. Each entry corresponds to a chemical element, a nucleus of a given electrical charge determined by the number of protons. Elements in a given column have similar structures in their outermost layers of electrons and therefore have similar chemical properties. The structure of this chart is determined by the quantum-mechanical, electromagnetic interactions of electrons with nuclei and with each other. The properties at this scale determine the potentialities available for larger-scale things such as molecules and living cells. Note that the atomic masses are not whole numbers. This is because they reflect the naturally occurring mixtures of isotopes, which depend on accidents of cosmic history.

machinery of life, including molecules that act as miniature computers and robots to encode the information in the machinery, others that decode and actually build it, and others that act as sources of energy to power it all.

Of course, this is just the nuts and bolts; it doesn't explain the miracle of how these structures get assembled, which is historical and therefore derived from cosmology. Physics determines what can happen, but it does not entirely determine what is actually there.

4 The Cosmic Expansion

he overall motion of the universe is expansion. If you take the large view, the universe is made of groups of galaxies all flying away from each other. Galaxies are big things, typically containing more than a billion stars each, but they are nevertheless much smaller than the observable universe (which contains about a hundred billion galaxies that we can observe), so it makes sense to picture the universe as an expanding cloud of galaxies. This idea lies at the heart of the evolving Big Bang model.

IS EVERYTHING EXPANDING?

The idea that the universe as a whole is expanding seems odd at first, because we are used to the universe on small scales, arranged into more or less stable assemblies of matter such as planets, stars, and galaxies. Matter is not expanding—you and I are stable assemblies of atoms. The solar system is not expanding but is a (nearly) stable arrangement of planets held in their orbits by the sun's gravity; the planets move, but in more or less the same paths year in and year out. Similarly, the galaxies are stable gravitating assemblies of stars and other matter. To some degree, even the galaxies themselves orbit each other in groups and clusters. But on very large scales—imagine a very blurry movie where you can make out only the average overall motion of large numbers of galaxies—the cloud of galaxies is flying apart, every piece moving away from every other.

This motion is related to the near uniformity of matter on large scales and to the fact that all directions in space are equivalent. The expanding motion goes along with uniformity: A uniformly expanding system can move and stay uniform (although it thins out with time) without choosing a particular direction. Rotation and shear, by contrast, pick out a direction in space, and nonuniform compression and rarefaction generate departures from uniform density.

The expansion is thus a signature of the simplicity of the universe on large scales and early times; it is the simplest way a universe can behave. A long time ago, in the early Big Bang, the universe was uniform on much smaller scales, and even very small bits of it were flying apart; *very* early, even things a few inches apart were flying away from each other. Today, matter on small scales has congealed into stable systems that no longer expand, because over small regions, where the expansion is not too fast, forces have reversed the expansion. On these small scales, things are no longer uniform; matter is in stable "lumps" (galaxies and their contents), which are flying apart from each other but are themselves not expanding.

ARE WE AT THE CENTER OF THE UNIVERSE?

It indeed looks as though we are at the center of the universe, but then it would look that way from any galaxy, anywhere and any time. Paradoxically, the center is both everywhere and nowhere.

Space is three-dimensional, but it is easiest to visualize the expansion in two-dimensional or even one-dimensional space. For example, think of a stretching rubber band with beads on it representing galaxies. All of the beads see all of the other beads moving away, yet none of them is in the center. The rubber band can keep stretching indefinitely, and every point can equally well be regarded as its center.[1] Also note that the farther two beads are from each other, the faster they move apart.

---

[1] It is not fair to cite the "center" of a circular rubber band, because this point is outside the band—that is, outside the universe. In the one-dimensional example, you have to imagine that the beads can look only along the band, not outside it.

Figure 4 shows a similar situation, but here the expansion is illustrated in two dimensions. A pattern of dots is shown at two different times; the patterns are identical except that the later one is

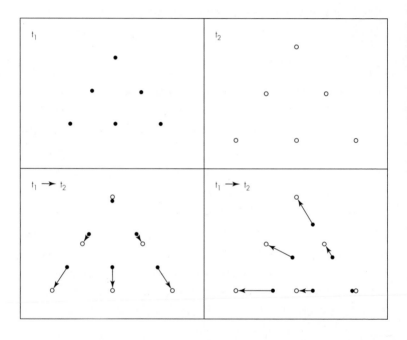

FIGURE 4 Expansion of a two-dimensional universe. The set of dots labeled t_2 is a slightly magnified version of the set labeled t_1 and represents the same region of the universe at a later time. You can line up any dot with its counterpart, and the others all seem to fly away from that point. The farther away they are, the farther they move in the time $t_2 - t_1$.

slightly enlarged. When you overlay the patterns, the system has a definite apparent center, from which all the other dots are receding. But you can also align the two patterns centered on a different dot. When you do this, you find that the expansion appears centered on whatever dot you have chosen. The patterns have not changed. What has changed is your interpretation of what has moved and what hasn't. Here, as in the rubber band model, between time 1 and time 2 the more distant dots have moved farther, which means that they are moving apart faster.

The real universe is just like these examples, with yet one more dimension added. A thin slice through space in any direction would reveal a situation like Figure 4; a long, thin tube through the universe would reveal motion like that of a stretching rubber band. Just because everything is flying away from us doesn't mean we are at the center. Rather, every point is the center—or no point is, because there is no preferred center. Any point will see the universe expanding around itself and can think of itself as the center.

WHAT IS THE UNIVERSE EXPANDING INTO?

Does the universe have an edge? Maybe, but not necessarily. It is possible for the universe to keep expanding, for all of space to keep getting bigger, without being embedded in some larger space. The universe may go on infinitely in all directions, or it may close back on itself, curving like a rubber band or the surface of a balloon.[2] In either case, the expansion has meaning only as viewed from within—there is no place "outside" from which to view it.

Alternatively, it is possible that there *is* an edge, but it must be far beyond the part of the universe we can see. We have seen no sign of an edge, or indeed of any real change from the way things are in our local part of the universe. If the Big Bang is like a stick of dynamite going off, everything we can see is within the expanding dynamite itself.

The *observable* universe has an edge, however, because the finite age of the universe allows us to see only a certain distance into space. As we look out into space, we also look back in time by the amount of time it takes light to travel that distance—a couple of seconds to the moon, about eight minutes to the sun, years or thousands of years to stars in our galaxy, about two million years to

---

[2]In this case there might be another, invisible, fourth space dimension (not the same as the fourth dimension of time that we already know about) through which the three-dimensional space curves. However, this extra dimension is no more observable than the third dimension of normal space is to inhabitants trapped on the curved surface of the balloon. In this "closed" universe, if you travel far enough in any direction in a straight line, you come back to where you started (but from behind), without ever encountering an edge. If the radius of curvature is large enough, it is hard to observe the global curvature for the same reason that a small piece of a large sphere is almost flat. It is also possible to have a three-dimensional hypersphere with uniform negative curvature. This "open" geometry has no two-dimensional analog, and may be infinite.

the farthest thing you can see without a telescope (M31, the Andromeda Galaxy); and so on. The most distant thing we can see by any means, even in principle, emitted the light we detect now shortly after the Big Bang, and this light has been traveling toward us for almost the whole history of the universe. We can never see farther than this, or rather, we will have to wait to see farther, about a year for every light-year farther away. Because the universe is only about 12 to 15 billion years old, allowing for the fact that the universe has been expanding, the most distant things we can see, even in principle, are now less than about 20 or 30 billion light-years away. In this sense, the observable universe does have an edge, but it is an edge in time rather than space; for example, in the deepest images we now see directly the formation of some of the earliest galaxies. It is remarkable that over this huge volume that we can see, every place is pretty similar to every other place—it's full of galaxies, everywhere—without any sign of an edge.

What Happened Before the Big Bang?

There are two parts to the question of what happened before the Big Bang: a mechanistic one about the expanding universe, another about time itself. Knowledge of the earliest moments of the Big Bang is not uniquely or precisely defined, partly because we do not know all of the physics involved, including the physics of time.

Why *was* there a Big Bang? One hypothesis (that of the "Inflationary Universe," described in more detail later) supposes that the physical vacuum of empty space developed an intensely repulsive gravitational interaction. This repulsion blew the energy of the universe apart, an explosion that started the Big Bang. According to this idea, the thing that started the Big Bang—really all that was needed to get a large expanding universe going—was a microscopic speck of excited vacuum. Given this speck, the Big Bang had to happen because of the forces of physics.

Many of the ideas of inflation are familiar and well accepted. The idea of repulsive vacuum gravity goes back to Einstein's early papers on cosmology; its connection to the actual cosmic expansion was made in the 1930s (almost as soon as the expansion was known) by Sir Arthur Eddington, an early champion of Einstein's theory of gravity; and the idea of an excited vacuum fits well into modern particle physics (indeed, this fit is the heart of inflation theory). But in spite of many modern ideas about how inflation might work in detail, there is no standard model of where the first quantum speck came from. This is hardly surprising, and it hardly matters from a physical point of view, because nearly all information is lost from that time anyway; it really was just a speck, of little interest except that it happened to be the beginning of the universe.[3] In other

---

[3]Actually, the fact that it is of "little interest" is itself a significant piece of information: If the speck had been too nonuniform, the universe today would not

words, the universe made its own information, it *made itself interesting*, so it is silly to worry at length about the initial speck as though it could take credit for everything that has happened since. If the speck is almost irrelevant and is impossible to find out about anyway, we should stop worrying about it and move on.

Did time itself have a beginning? Contrary to our intuition (which is probably based on our experience that one time is much like any other), time as we know it may not be infinite in the direction of the past. As with space, this does not necessarily mean there is an edge (an earliest moment before which there was nothing). Rather, it is perfectly consistent to imagine that as one goes far enough back, the notions of "before" and "after" become indistinct, and time ceases to have the precise meaning it has in familiar physics. We already have established the inconsistency of extrapolating the conventional notion of time to very small intervals, so when the beginning is approached too closely, we cannot use the concept of time. "Time" and "space" are both concepts that makes sense only for a system that has a certain minimum size and duration; there are ideas for descriptions of smaller or shorter things (whatever that means), but they have not yet been tested against reality.

---

be as uniform as it is observed to be even after the smoothing effects of inflation. The current idea is that all of the nonuniformities are due to tiny random quantum fluctuations (see Chapter 8).

Compare this question with medieval speculations about what happens at the edge of the world. A believer in a flat Earth is faced with either an infinite world or one with an edge, whereas with a round Earth, the question of an edge ceases to have any relevance. Asking what came before the Big Bang might be like asking what is north of the North Pole—a place where "north" has no meaning. It is presumptuous to assume that just because they are suitable for talking about the nature of time today, our ideas of time must also apply to the utmost extremities of spacetime.

The structure of a four-dimensional spacetime can be visualized if we are willing to drop some dimensions. Figure 5 shows a partial view of a complete spacetime illustrating the expansion of space, its curvature, and the path of light travelling through the expanding curved space along this surface.

In this picture the beginning appears as a point since things were at first packed closely together. Horizontal sections above this reveal ever larger circles, each of which represents the

FIGURE 5 An expanding universe spacetime. The surface represents the entire past history of all of space. Time runs vertically, the Big Bang at the bottom and the future at the top. Space runs horizontally; as the universe expands, the spatial sections get wider. Only one space dimension is shown here, (it would not be possible to show all three), and this particular figure shows a particular model of space with spatial sections ("hyperspheres") that close back on themselves. An open universe expands similarly (and looks the same locally) but does not recollapse at the top or curve back on itself like this. A single galaxy's timeline extends along a

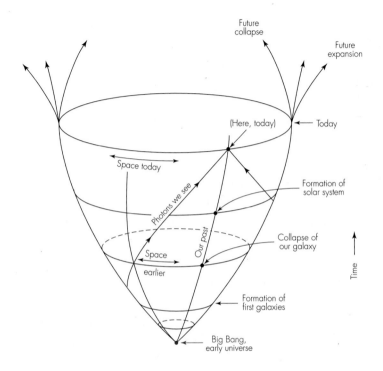

vertical stripe, and photons move diagonally; indeed, a small piece of this spacetime looks exactly like the local spacetime diagram (Figure 1, p. 27). Our location is shown now and at the formation of the solar system and of the galaxy; the region probed by our telescopes (which look along past photon trajectories) is also indicated. Light coming from objects at large distances carries information about the global curvature of space and the history of the expansion. The beginning of time is at the lower tip of the football. It might be like the North Pole—In particular, the concepts of "before" and "north," respectively, lose their usefulness at these places.

whole of space at various times (in a particular model), not showing two of the three dimensions. The scale is of course enormous—slices today have a circumference of at least tens of billions of light years.

Figure 6 takes the expansion out and shows a different slice through the spacetime, the present-day positions of things seen at various times in the past; this is a way of showing our past "light cone," the events we see by looking out into space and back in time. Each horizontal section here represents events on a great circle around the sky. This view better illustrates the "edge" created by looking back to the beginning; the distance to this cosmic horizon is again tens of billions of light years.

Although these figures are only sketches of possible spacetimes, we can map the global structure of the actual universe by observing light that has travelled from distant objects.

It's also possible that the "initial speck" was part of another, larger universe. In that case, time might go back forever, and might even be embedded in other universes, but we are unlikely ever to get any information about the "pre-speck" part—so who cares?

Most of these issues don't really matter when we are discussing the Big Bang model and its predictions. Inflationary models share the property that if there was anything "before" the Big Bang, the evidence of it was long ago erased or diluted. The large-scale structure we see today probably reflects the inflationary period and events of the more recent universe; the composition of

Chapter 4 The Cosmic Expansion

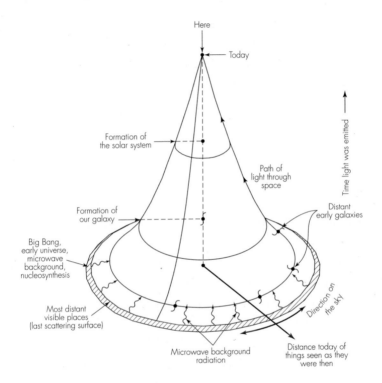

FIGURE 6 A view of events on a single plane of space on our past lightcone. Here two spatial dimensions are shown (each direction around the circle corresponding to a direction on a great circle on the sky), and time is plotted vertically. The effects of the expansion have been removed—the cone shows the present-day spatial positions of points whose light we are receiving now. We are at the apex of the cone and see a larger volume of space as we look back in time; however, the volume we can see is not infinite. The height of the cone (the age of the universe) is about 13 Gy; the radius (the size of the observable universe) is about 20 or 30 billion light-years.

the Universe reflects the thermal cauldron of the early Big Bang after inflation; and the complexity of structure on larger scales, from molecules to galaxies, reflects billions of years of cosmic evolution. These are things we can study scientifically, by observation and modeling.

Effects of Expansion on Light

Light traveling through the expanding universe notices the expansion. The wavelengths of light from distant galaxies have been stretched by the expansion from what they were originally, in proportion to the overall size of the universe.[4] A bit of blue light after a while (after a 20 percent expansion) becomes yellow, and later still (after another 20 percent) it becomes red. After a very long while, a thousandfold more expansion, it becomes microwaves, and after another factor of a hundred, radio waves. The expansion's effect on primordial light is to "cool things down" with time. The

---

[4]This notion of "stretching" is the most accurate verbal metaphor for the mathematical description of the cosmic redshift. Over short distances (such that the expansion speeds are much smaller than the speed of light), an equivalent description is the Doppler shift, the redshifting of light due to the expansion velocity. This description is however both harder to explain and less scrupulously accurate.

universe started very hot, filled with energetic, short-wavelength radiation that has now become cool, low-energy, mostly microwave radiation. The effect on light from galaxies is also noticed: Light from more distant galaxies, which was emitted a longer time ago, has experienced more of this stretching (or "redshifting"), which indicates that the universe was smaller in the past; all the galaxies were closer together. (See Figure 7.)

Wavelengths of light propagating in an expanding universe increase in proportion to the size of the universe. The size of the universe when light left an object can be identified by the amount of stretching of identifiable, particular wavelengths of light that originated as "pure" colors, belonging to the "favorite" wavelengths of the atoms in the stars and gas of the galaxy that emitted the light. One familiar pure color is the intense red hue of the lasers in a CD player or supermarket checkstand. This color is always exactly the

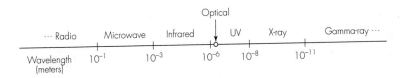

FIGURE 7 The electromagnetic spectrum, showing the names given to different wavelengths of light. The expansion of the universe causes a "red shift," a change of color from shorter to longer wavelengths.

same, because the lasers are made of the same material, so if you saw one with a different (redder) hue, you would know it was redshifted.[5] Once such a color is identified in a distant galaxy, its known original wavelength can be compared with that seen, and this comparison reveals how much the universe expanded while the light was en route to us. The most distant objects therefore have the highest redshifts—their light is the most stretched out.

The brightest things in the universe, and hence the objects most easily studied at high redshift, are energetic sources called quasars in the centers of galaxies; in the most distant one yet found, light we see now has been stretched to nearly six times its original wavelength, having left the quasar when the universe was about one-sixth as large as it is today. Light from earlier times has stretched much more than this. The light from the Big Bang itself originated much earlier, before weak decoupling, when the universe was less than a ten-billionth as large as it is today and more than ten billion times hotter.

---

[5] The color, of course, also tells you about the material. In astronomy, for example, lines of hydrogen are very common because it is the most abundant element. Indeed, the composition of the universe as a whole is deduced from just such clues. For instance, helium is named after the sun because its existence was first noticed in the colors of light coming from the sun! Usually a given substance has more than one favorite color, so we can use the patterns of colors to tell the difference between the effects of cosmic redshift and those of composition.

THE HUBBLE LAW AND THE EXPANSION RATE

This stretching of light can also be thought of as a redshift caused by the motion of matter, in the same way that a siren seems to decline in pitch on a fleeing ambulance and to rise on an approaching one. For light, the redshift (the fractional change in wavelength) is the velocity divided by the speed of light. If the expansion velocity is not too large, then the cosmic stretching of wavelengths is equivalent to the famous "Hubble law" describing the motion of uniformly expanding matter,

$$\text{Velocity} = H_0 \times \text{distance}$$

which relates the relative velocity and distance of any two galaxies. As our rubber band example showed, the farther they are apart, the faster they move apart. The Hubble Law is a good description of the real universe, as shown in a Hubble Diagram of relative distance versus velocity or redshift on large scales (see Fig. 8).

There is a number H_0 in this formula, called the Hubble constant, which we have to find by measuring not relative but absolute distances. If the Hubble Law were exact, we could measure the Hubble constant by finding the velocity (or redshift) and distance of any galaxy. Velocities are measured accurately by the shift of the wavelengths of light. Distances are also measured accurately if they are small enough, by using triangulation or parallax (see Figure 9),

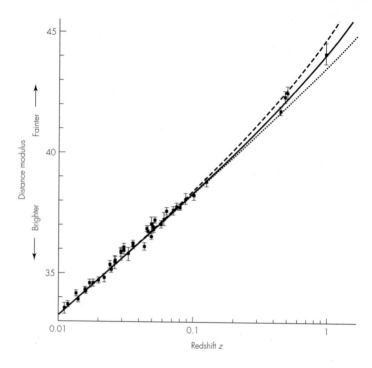

FIGURE 8 A Hubble diagram of Type Ia supernovae. A Type Ia supernova is a type of exploding star whose brightness is a good indicator of distance: fainter ones are farther away and receding faster. The redshift z of each supernova is plotted on the horizontal, where $1 + z$ is the factor by which its light has stretched; for z much less than 1, the expansion velocity is z times the speed of light. On the vertical the brightness of each is shown in units ("distance modulus") where 5 units correspond to a factor of 100 in brightness or factor of 10 in distance. The relation of distance to velocity is close to the linear relation predicted by Hubble's law for a uniformly expanding universe. Possible deviations from this law as z approaches 1 can arise from changes in the expansion rate or from the global curvature of space; the curves show some plausible options.

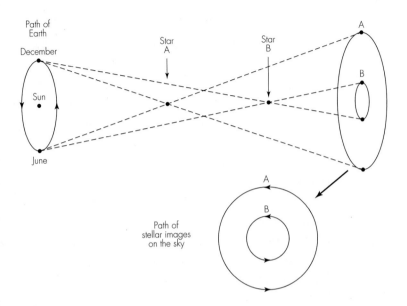

FIGURE 9 Using parallax or triangulation to measure small cosmic distances. As Earth moves around the sun in the course of a year, nearby stars appear to shift their position in the sky relative to more distant ones. The angular motion is larger for nearby stars and can be used to estimate their distances. (The stars in the figure are much closer, and the angles much bigger, than for the nearest real stars.)

the same technique used by surveyors. We can measure the distance to nearby stars by using Earth's orbit as a baseline; the distance of a star is found by measuring the tiny change in angular position of a star at different times of the year relative to very distant stars. This ties small cosmic distances to local meter sticks. But large distances can't be measured this way (the angles are too small), and because the Hubble relation applies only to a large-scale average motion of galaxies, measuring H_0 requires measuring large distances accurately, a technically challenging problem. This is why the value of H_0 is still uncertain by about 20 percent even though the large-scale validity of the linear Hubble relation is more precisely verified.

The value of the Hubble constant is interesting because it tells how quickly things are flying apart. Because things move faster in direct proportion to distance, any galaxy traverses its distance to any other in the same time, regardless of their separation. A single number, the inverse Hubble constant $1/H_0$, thus indicates how long ago all galaxies were in the "same place," assuming that the expansion has always had the same speed.[6] Current estimates place $1/H_0$ at around 14 to 17 billion years. If galaxies had never changed speed, this is how long ago they would have had zero separation—the elapsed time since the Big Bang occurred.

---

[6]They were never actually in the same place; galaxies did not yet exist in the early universe. But it is true that all the matter now in our galaxy and all the matter now in other nearby galaxies were once packed into a volume much smaller than that which the solar system occupies now.

CHANGES IN THE EXPANSION RATE

But galaxies, and the pregalactic matter before them, have not always been flying apart at the same speed. The expansion rate has been changed by gravity.

If you throw a baseball straight up into the air, it will fall back to Earth—unless you throw it very hard indeed, in which case it will slow down as it rises but will never turn around, flying away from Earth forever. These same options exist for the expansion of the universe; in fact, precisely the same mathematical equation describes the motion of a baseball in free-fall flight and the separation of any two galaxies in the expanding universe. The physics is the same—in both cases, the motion is controlled just by the force of gravity. The only difference is that the baseball's motion is controlled by the gravity from the mass of Earth, whereas the galaxies' motion is controlled by the gravity from the mass of galaxies themselves and other mass of the universe.

Corresponding to the slowly thrown baseball is the slowly expanding, dense universe.[7] A dense universe slows down and eventually stops expanding; then it contracts, as all the galaxies rush

---

[7]The expansion behavior is connected to space curvature. High density curves space globally; gravity not only influences the expansion but also reflects the global space curvature. Universe that recollapse also tend to have hypersphere geometries which close back on themselves.

together with increasing speed, ending in a "Big Crunch." The fast, escaping baseball corresponds to the rapidly expanding, low-density universe that flies apart forever, the galaxies thinning out so much that gravity ceases to have much effect. As such a universe gets very old, the galaxies eventually hardly slow at all and never even get close to zero velocity, instead coasting at some limiting speed. Between these two cases lies the delicate balance, a universe that is like a baseball thrown at exactly the escape velocity from Earth—just the right balance of expansion speed and density so that it eternally approaches, but never attains, zero velocity.

With the universe as with the baseball, it is hard to tell which of these cases applies just by looking at the situation at a given instant. If you throw the baseball up just a little, it will fall back right away, but if you throw it really hard, it will fly far into space before returning—and it is difficult to tell whether it will ever come back. A universe can similarly go for a very long time before it finally, if ever, turns around (and indeed this is the case for our universe; it will have lasted for many tens of billions of years before, if ever, it turns around).

The escape velocity for a baseball depends on the mass of Earth and also on the exact height from which it is thrown. Similarly, the fate of the universe depends on exactly how fast it is expanding and on how much mass is slowing it down. These things have not yet been measured precisely enough in the real universe for us to know what is going to happen. The best way to tell is to compare the expansion rate in the past (by looking very far away) with that in the present to see whether the rate is slowing. This effect has not yet been reliably measured, although we have promising techniques (Figure 8).

The universe, unlike the baseball, has another option: to speed up as it flies apart, like a ball in a slingshot or rocket. This is possible because in principle, gravity can actually act as a repulsive force on empty space. Einstein introduced this possibility of a "cosmological constant" shortly after his theory of gravity; it occurs if empty space devoid of all matter and radiation nevertheless contains energy and couples to gravity. If the vacuum has enough energy in it, the force driving the universe apart triumphs over the attraction of the matter gravity, and the expansion gets faster and faster with time. This possibility has been known ever since the expansion was discovered, but we still don't know whether the actual universe is speeding up or slowing down. Measuring dynamics on cosmic scales seems to be the only practical way to measure the energy of the vacuum.

THE AGE OF THE UNIVERSE

Even after we know the Hubble constant, or Hubble time, $1/H_0$—the amount of time it would take galaxies to get where they are from one point, traveling at their present speeds—we will not know the actual age of the universe. This depends on how the expansion has changed with time, which depends on the kind and amount of matter in the universe. But we can turn this around: If we can measure the age of the universe in some other way, comparing it with $1/H_0$ will tell us something about the kind of universe we are living in and what its future might be.

We have a pretty accurate idea of how old the solar system is: The sun, Earth, and the other planets formed from interstellar gas about 4.55 billion years ago. This estimate is based on the physics of radioactive decay. Certain atomic nuclei, such as uranium-238,[8] are unstable and decay into other nuclei. They do this at a fixed rate that is well measured. For example, after 4.6 billion years, a single uranium-238 atom has a 50 percent chance of converting to lead-206, and after this time has elapsed, an initially pure lump of uranium-238 is half lead-206. Thus if we can find naturally occurring samples of rocks and meteorites containing deposits of uranium that started without lead-206 (or had an initial amount that we can guess from other isotopes of lead), then the ratio of lead-206 to remaining uranium-238 tells us how long ago the deposit formed. Many objects in the solar system, especially a wide variety of meteorites, have been measured using different radioactive nu-

---

[8]Recall that there are different uranium nuclei, depending on the number of neutrons; 238 is the total number of neutrons and protons in this particular type of uranium nucleus. All uranium nuclei have 92 protons.

clei (for example, potassium-40, which decays into argon-40 with a half-life of 1.3 billion years). All give the same age of 4.55 billion years, so we know the universe is at least this old.[9]

Although stars are forming in the universe today, many stars are much older than the solar system. Some are two or three times as old, and indeed it is likely that some of the first stars to form after the Big Bang are still burning today. We can use these old stars to estimate the age of the universe.

It would be very helpful if someone had started a stopwatch at the instant of the Big Bang and had left it around so we could just read off the age. In fact, we have something quite like this: old clusters of stars with a variety of masses that function as sets of differently sized hourglasses running at different rates. If we can somehow learn to read the stars as hourglasses—which means calibrating stars by using our understanding of how they work—we can read off the age of the star clusters by reading off the age of the hourglasses about to run out. To do this, we need to apply a model of how stars work.

Stars like our sun, which still have plenty of fuel left, remain in a stable state for billions of years. They are balls of gas, mostly hydrogen nuclei and free electrons, held together by gravity and held

---

[9]A similar exercise can be applied to whole stars: The amount of radioactive thorium in old stars, relative to the other elements that formed alongside it, is less than in more recently formed stars. This is not, however, the most precise way to tell the ages of stars. For this we examine the lifespan of the stars themselves.

apart by the flow of heat from the nuclear reactions in their interiors. Most stars spend most of their lives in this stable "main sequence" state; most of the stars you see in the sky are like this.

This stable state eventually ends, however. The self-regulating balance between the forces is upset when the hydrogen fuel in the center begins to run out. Lacking fuel, the central core collapses. This releases more heat than is required to hold up the outer parts of the star and drives off the outer layers. Thus the structure of the star changes when it runs out of hydrogen in the core; the inner part becomes smaller, the outer layers larger. We can see the effects of these changes on the light coming from the star. Because the light is coming from a larger area, it is more spread out, which also changes its temperature—the star becomes cooler, and its light becomes redder.

Stars are like hourglasses in that once they form, they start using up their initial allocation of fuel (hydrogen gas), and when they begin to run out of fuel, they change color. This color change signals that the hourglass has run out.

More massive stars are brighter than less massive ones and use up their fuel more quickly, acting as faster hourglasses. Therefore, if a star cluster forms with a variety of masses, the most massive stars exhaust their fuel first, followed by successively less massive ones. Measuring the masses of those that are just now running out of fuel gives us the age of the cluster.

We can use stars as clocks because we understand what goes on inside them as they evolve. Models predict, for each mass and composition of star, its observable properties (brightness and color) at each time in its life; the measured brightness and color then indicate the time elapsed since formation (see Figure 10). The time accuracy of the models is confirmed with other tests. The predicted age of the sun, for example, agrees with the radioactive dating of the solar system. And where the masses of stars are precisely known, brightness and color are correctly predicted. There is even a case where two stars orbiting each other, with different masses, are measured to have the same age.

The current estimates, derived with these techniques, for the ages of the oldest star clusters are around 12 to 15 billion years. The oldest clusters seem to have ages within a few billion years of each other, and nothing much older than this has been found—a remarkable fact that suggests the universe is about this old. Even more remarkable is the agreement with the "expansion age" of about 14 to 17 billion years derived from the Hubble constant with no reference to local clocks. We now have a fairly secure, if still imprecise account of cosmic chronology (see Fig. 11).

When we look in detail at the numbers, it seems that in some models of the expansion—the recollapsing ones, for example, moving at less than the escape velocity, where there is continuous gravitational slowing—the stars are older than the universe. The predicted age in any recollapsing model is at most $(2/3)/H_0$, or about ten billion years. These are fairly "young" models because the galaxies were flying apart faster in the past than they are today.

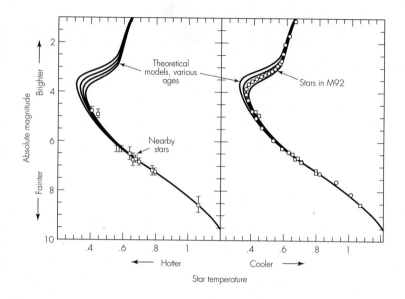

FIGURE 10 The absolute brightness of stars (in astronomical magnitudes where five units correspond to a factor of 100 in light output), plotted against their temperature as measured by an index of color, with hot to the left. The left figure shows some nearby stars, the right figure stars in a globular cluster in our galaxy, M92. The dots represent the observed stars in this cluster. The lines represent theoretical models of where they ought to lie at several different ages for the cluster, ranging here from 12 to 18 Gy. For these models, it is assumed that stars of different masses formed at the same time, and a model of their evolution is used to calculate their present appearance. The models bend at the stars (just under a solar mass) where the fuel is just now running out. Stars that were originally in the cluster and were more massive than these have used up their core hydrogen and have left the stable main sequence. The best-fitting model gives the age of the cluster, in this case about 15 billion years.

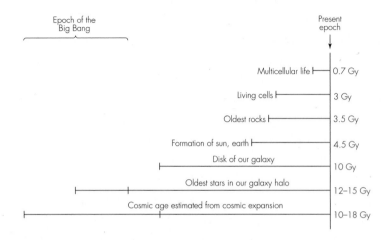

FIGURE 11 The age of the universe and its constituents. Although the numbers are not precisely known, we do know that our galaxy is not many times older than Earth and the sun and that the universe is only a little older than the galaxy. We find rocks on the surface of Earth today (even rocks with fossils of life) that are more than three billion years old—about a fifth of the age of the entire universe!

They got to where they are in less time, predicting a universe that is too young.[10]

_____

[10]Note that ruling out these models does not rule out the Big Bang, as is sometimes implied in the popular press. It seems strange at first that we can know a lot about the early universe when some basic facts about its later evolution are still unknown, but this is indeed the case. The behavior of the universe at early times depends rather little on the eventual fate, because at early times all the models are expanding at very similar rates.

The universe can be old if the matter density is low, so that there has been little gravitational slowing of the expansion in the last ten billion years or so; alternatively, if there has been some acceleration from a cosmological constant, then they were traveling even more slowly in the past, and the universe can be older still. Either way, it appears from the present evidence that there is not enough gravitational slowing to make the expansion reverse itself. The universe will probably expand for much longer than it has already, and possibly it will expand forever.

The mere fact of the expansion of the universe implies that matter was once very much closer together than it is today. The recent behavior of the expansion does not, however, tell us much about the behavior of the universe during the first billion years, the first million years, or the first minutes or seconds. The most important of clues to these periods of cosmic history are the background radiation and the composition of matter, which we shall consider in the next two chapters.

5 Cosmic Background Radiation

Space is not dark and empty but filled with radiation. Most of its energy lies in the microwave region of the spectrum, near wavelengths of about 1 or 2 millimeters, coming evenly from all directions. This light, called the "cosmic microwave background radiation," is left over from the Big Bang. It has been moving freely through space since the universe was less than a thousandth of its present size and much less than a million years old. The light comes directly from the time the universe first became transparent to light, so it is the most distant thing as well as the earliest epoch that can ever be imaged directly with light. Its properties reflect the simplicity of the universe at early times and on large scales.

THE GLOWING SKY

The first evidence that the microwave background radiation is coming from the distant universe is that, as predicted for the radiation left over from a uniform Big Bang, the radiation intensity is nearly the same in all directions. This remarkable uniformity is unique: Nothing within the universe is as uniform as the universe itself.

The night sky viewed at microwave wavelengths shows only a trace of "foreground" emission from our galaxy, a strip corresponding to the Milky Way Galaxy. Most of the radiation is background light which glows quite evenly from all directions, like the sky on a bright, clear day. This light is coming from beyond all the galaxies.

What are we actually seeing when we look at microwave background radiation? It is similar to looking at the surface of the sun. There is no hard surface to the sun; it is a ball of gas that diminishes in density and temperature gradually toward the outside. What looks like a surface or an edge is a region within which the gas gets so hot that it is no longer transparent to light. This region (called the photosphere) appears narrow because of an abrupt increase in the interaction of matter and radiation at a certain temperature. The sudden transition creates what resembles a surface, which makes the sun look like it has a certain size.

From the photon's point of view, the "surface" is the point where it can stop scattering off of matter all the time and travel freely out into space. Similarly, when the universe cools to about 3000 K,[1] it becomes transparent. The electrons in the hydrogen gas (the main cause of the scattering of light in this situation) stick to protons and become neutral atoms, which (like the gas in our atmosphere) is transparent, so you can see right through it. After that the photons of the background radiation fly straight to us without further interference. This cosmic photosphere is very much like the surface of the sun—it is made of similar materials, at a similar temperature—only we are on the inside looking out.

Like matter, radiation has a temperature. The background radiation used to be hot, but now it is cool; a thermometer placed in deep space today will read 2.726 K above absolute zero, 1100 times cooler than it was at last scattering. The low temperature—a result of the universe's having expanded for a long time—is a measure of just how dark the sky is at night. The sky is not actually dark; it is just like the surface of the sun, only 2000 times cooler. The near uniformity of the background intensity corresponds to a near uniformity of temperature, reflecting a nearly identical early cosmic history and uniform expansion in all directions.

_____

[1]Note that the "cosmic photosphere" at 3000 K is a little cooler than the solar photosphere at 5800 K, because the matter of the universe at the same temperature is much less dense than that of the sun.

OUR REFLECTED MOTION: THE COSMIC DIPOLE

A detailed examination, however, reveals a very slight variation—
about 1 part in a 1000—in the temperature of the light across the
sky. One direction is a little hotter than average, the opposite direction a little cooler, and in between there is a remarkably smooth and
gradual transition (Figure 12, top). This variation shows that we are
not at rest in the universe but are moving through it.

This is to be expected as things are not expanding on small
scales. Earth goes around the sun, the sun goes around the galaxy,
and the galaxy is falling toward its neighbors. All of these motions

---

FIGURE 12 Sky maps of the microwave background, from the Cosmic
Background Explorer satellite. Each map is an image of the entire sky in microwave light, showing the variations in temperature or intensity of the radiation. At these wavelengths, the sky is dominated by the cosmic background
radiation, so these maps are direct pictures of the early universe, with only a
small amount of foreground emission. Different shades correspond to small
variations in temperature in different directions. The top frame shows the dipole anisotropy of the microwave background—the gradual variation across
the sky from one side to the other, caused by the motion of Earth through
the universe. The uniform radiation appears slightly hotter (slightly more intense, and slightly shorter wavelengths on average) in the forward direction
and slightly cooler behind. When this effect is subtracted (center frame), the
Galaxy shows as a trace of contamination across the center, corresponding to
still smaller variations. When this too is subtracted using a model of the
Galaxy, the radiation is smoothly coming from all directions, except for
barely detectable fluctuations of a few parts in a hundred thousand (bottom
frame). These appear to belong to the universe itself and contain information
about the origin of cosmic structure.

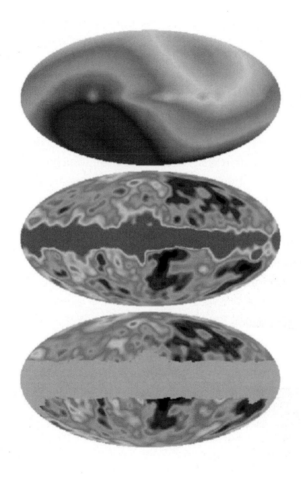

sum to a velocity of about one thousandth of the speed of light, typical of departures of galaxies in the universe from uniform expansion. If we were not moving relative to the "cosmic frame," the background would appear the same in all directions to much higher precision.

Maps of the cosmic background radiation tell our velocity through the universe quite precisely—both how fast we are going and in what direction—to an accuracy of a few percent, better than a car speedometer. It is significant that there is such a thing as a velocity through the universe, because absolute velocity is not defined by the laws of physics per se, which recognize only relative velocities; indeed, this "principle of relativity" is a cornerstone of our theory of spacetime. However, the distribution of the matter and radiation of the universe does define a "preferred frame" that, at each position, is the frame in which the universe looks the same in all directions. The solar system is moving relative to that frame, with a velocity of 370 kilometers per second (Figure 13), toward the head of the constellation Virgo.[2]

---

[2]The actual coordinates are right ascension 11 hours 12 minutes, declination −7 degrees, which lies near the boundary of the constellations Virgo, Leo, and Crater. The observed anisotropy depends slightly on the season, because Earth circles the sun once a year with a velocity of about 31 km/sec. The sun is also orbiting in the galaxy at about 220 km/sec, but this takes 300 million years for one orbit, so we don't detect any change in direction. The galaxy itself is moving rather fast relative to the cosmic frame, about 600 km/sec. All these motions of the solar system add up to cause the dipole anisotropy (see Figure 12).

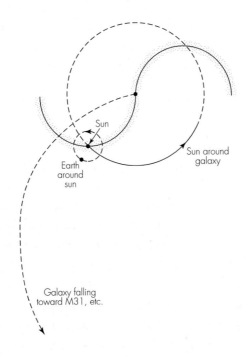

FIGURE 13 The various motions of Earth relative to the local cosmic frame of reference, and part of the hierarchy of systems it belongs to. When all are added together, we are moving at about 370 kilometers per second relative to the local cosmic frame.

Because 370 km/sec is 0.0012 times the speed of light, there is a very gradual shift in temperature, by 1.2 parts in 1000, from one side of the sky to its opposite. When the gradual variation due

83

to this motion is subtracted and the residue from the galaxy is subtracted, the remaining radiation looks amazingly uniform—to a few parts in 100,000. This is much smoother than a billiard ball; a freshly polished ice rink may come close to the same fractional variation.

INTRINSIC COSMIC FLUCTUATIONS

Those tiny variations that *are* seen in background radiation temperature, after subtracting the dipole, probably contain the key to the mystery of why, after hundreds of millions of years, cosmic structure finally developed (see chapter 7).

We may therefore already have maps of the sky showing large-scale fluctuations that give a clean and direct view of primordial inhomogeneity, shown by the wrinkles in temperature in Figure 12. Probably these are intrinsic fluctuations that were already present in the universe at recombination; the temperature fluctuations are imprinted by a combination of gravitational redshift, gravitational waves, matter motions, and actual temperature variations. It is also possible that the universe at recombination was smooth and that these wrinkles were imprinted afterward. Either way, the wrinkles are relics of whatever process caused the universe to break up into structure. The present data have ruled out many models of structure formation (for example, those based on chain reactions of star explosions) and suggest that the wrinkles are a relic of inflation. New maps of the radiation with much finer detail will soon be produced by new satellites; these will contain enough information for us to deduce the precise causes of the anisotropy and of cosmic structure, as well as the precise values of many cosmological parameters. Models predict that finer resolution will show somewhat larger-amplitude ripples on somewhat smaller angular scales, but very smooth structure on the smallest scales. The exact appearance of the map will depend on the new physics creating the fluctuations.

THE COLOR OF PRIMORDIAL LIGHT

The most spectacular and precise confirmation of the Big Bang is found in the spectrum of the cosmic background radiation—its color.

The spectrum of a light beam is its mixture of colors: the relative amount of energy of light of each wavelength. Sunlight passed through a prism or a water droplet reveals a rainbow, which shows that sunlight is really a mixture of different pure colors. The actual detailed mixture—the brightness of each color relative to the others—tells us about the temperature and composition of the sun.

Light always originates from some kind of emitting matter, and the spectrum of light always contains information about where it came from. Because all colors propagate in empty space at the same speed, the mixture of colors in a light beam does not change with time. Light traveling in empty space "remembers" its source.

Although the cosmic expansion changes the colors of light, it does so equally—all wavelengths shift in proportion to the size of the universe—so after allowing for the uniform stretching, the mixture of colors is unaffected by the expansion. Thus the spectrum of the microwave background "remembers" the early universe.

Light from the sun is not just a random mixture of colors but a particular mixture determined by the way the sun emits light. Amazingly, this mixture is the same as that of the primordial radiation, except that the primordial light is both cooler and purer. The spectrum of the light in the microwave background reveals its source as surely as the red of the laser, the green of grass, and the blue of the sky. It has what physicists call a blackbody spectrum (Figure 14). The mathematical form was written down at the turn of the last century, in Max Planck's blackbody radiation formula.[3]

---

[3]Planck's formula is

$$B_\nu = \frac{2h\nu^3}{c^2} \frac{1}{e^{h\nu/kT} - 1}$$

where B_ν (the spectrum) is the energy flux at frequency ν, h is Planck's constant (which connects frequency and energy, the first appearance of quantum theory in physics), c is the speed of light; k is Boltzmann's constant (which converts temperature units to energy, reflecting the fundamental thermodynamic result that the typical energy of each particle in a gas is about kT); and T is the temperature of the thing that is radiating the light. The data from the sky (figure 14) match this formula for $T = 2.726K$, the current cosmic temperature.

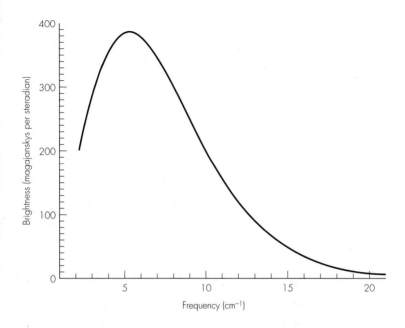

FIGURE 14 Spectrum of the cosmic microwave background, measured by the Cosmic Background Explorer satellite. Brightness (energy per solid angle per frequency interval) is plotted against frequency. The curve is a perfect blackbody spectrum; no deviations have been found (the experimental uncertainties in the measured sky spectrum are much smaller than the width of the printed line). It agrees exactly with Planck's formula to a few parts in 10,000. A light bulb and a stove top emit blackbody radiation of higher temperature. Their spectra, and those of all other opaque materials, are nearly identical in shape to this curve, except that the axis labels are different for higher temperatures, corresponding to shorter wavelength and higher intensity. The background radiation is the most perfect blackbody known; the limit on the precision of this measurement was the precision of the artificial blackbody reference source.

Mathematically, this spectrum arises because light comes in discrete "quanta," or packets, of energy (called photons). It appears whenever photons can be freely created and destroyed and can exchange energy with each other by interacting with matter so that their energies are randomized. This is an example of statistical physics: The randomization always turns up the same spectrum, just as surely as a coin will come up heads about half the time if you toss it many, many times. Electromagnetic fields that emerge as light act like connected springs being shaken (the effect of the temperature). The energy of the shaking will distribute itself in precisely a certain way between fast and slow shaking, the relative amounts being given by Planck's formula.

Planck's formula describes the mixture of colors coming from any opaque piece of matter at a single temperature. The spectrum depends on only one number, the amount of energy available for making photons, and this is determined by the temperature of the blackbody. The formula says that hotter things radiate more light than cool things do and that they also radiate mostly higher-energy photons, corresponding to shorter wavelengths of light.

Planck's spectrum is the mixture of light emitted by matter just by virtue of having a temperature and interacting with radiation. Although the specific temperature varies, this spectral form is ubiquitous in nature. It is the same for light from the sun or from a light bulb, for the infrared light from a stove top or a dying campfire, and for the still cooler invisible radiation you can feel coming from warm beach sand after sunset. The light from the sun is short-wavelength because of the high temperature at the surface, about 5800 K; most of the energy comes out at wavelengths we can see with our eyes, and some even comes out at higher ultraviolet (UV) energies that cause sunburn. The filament of a light bulb would vaporize at such a temperature; even the hottest halogen bulbs are cooler than the sun, so most of their light appears in photons of proportionately lower energy. Light bulbs therefore emit proportionately more invisible infrared light, which is felt as heat, and less ultraviolet light. The warm sand of the beach is several times cooler still, about 300 K, and its light is of still longer wavelength such that it can no longer pass easily back to space through Earth's atmosphere because of murkiness from carbon dioxide, methane, and water. (The fact that energy can get in more easily than out leads to the "greenhouse effect.") A temperature of 300 K corresponds to a light wavelength of about a hundredth of a millimeter, 20 times longer than visible light.

The cosmic background radiation energy arrives at even longer wavelengths, from about 5 to 0.5 mm. This is about 100 times longer than the light from beach sand, corresponding to a 100 times cooler temperature. Amazingly, its spectrum agrees to better than 1 a few parts in 10,000 with the precise formula that Planck wrote down in 1900 for blackbody radiation! In other words, with one number, the temperature, a simple mathematical formula gives the exact spectrum of light coming from all over the sky, to a precision of better than a tenth of a percent, over more than a factor of ten in wavelength. This simple but precise result was predicted by the Big Bang model before it was found in the sky. The actual temperature of the sky is 2.726 K, barely above absolute zero and about 2000 times cooler than the Sun, but in all other respects its spectrum is almost exactly the same. When the universe was 2000 times smaller than it is now, the light looked just like the surface of the sun, everywhere.[4]

Because the Planck spectrum crops up all over the place, it may not seem remarkable that it is coming from the sky as well, until we remember that the matter of the universe today is not a blackbody at all but is transparent; we see galaxies and quasars to enormous distances even with visible light, and microwaves are even more penetrating. The microwave background is coming from the dark sky between the stars and galaxies—from behind them. Somehow the universe back there makes a spectrum of light that looks exactly like starlight, only much cooler.

The blackbody spectrum tells us that the universe was not always transparent. To establish the Planck spectrum, light of many wavelengths must be coupled to opaque matter at a single temperature. At the observed wavelengths, the universe is transparent and

---

[4]The universe is actually a more perfect blackbody than the sun, which has imperfections from sunspots, storms, granulation, and absorption lines.

the radiation now travels freely through space, so interaction with matter cannot create a blackbody spectrum. In the Big Bang model, the blackbody spectrum is a relic of the earlier, hotter, denser state when the interaction of radiation with matter was more efficient. The universe is a blackbody for the same reason that starlight is, and the spectrum tells us directly that the universe has evolved from a hot, dense early state like the state of matter and radiation today inside stars such as the Sun. The precision of the blackbody spectrum now measured indicates that at least 99.9 percent of the energy we see was already present when the universe was a million times smaller and a million times hotter than it is today—a temperature of several million degrees, almost as hot as the center of the sun.

The entire expanding universe at that time was only a few weeks old, and most of its mass was still radiation and not matter. The background radiation thus directly verifies the heart of the Big Bang model; it carries a message directly from the first weeks of the expansion, and it reveals what, in the simplest Big Bang model, we expect to find. No other model, not even much more complicated ones, matches the data so precisely.[5]

---

[5]Although the radiation is not exactly uniform across the sky, the blackbody character is; the variations in intensity are caused by variations in temperature, preserving the blackbody spectral form, as predicted. Recall that the temperature variation due to Earth's motion is only 0.1 percent, and the intrinsic variations are much smaller still.

Although the cosmic background photons originated when the universe was a million times smaller than today, for a long time after that they continued to scatter off matter, changing their directions frequently. We thus view the early universe through a sort of fog, but like a traffic light hidden in fog, the color of the light is visible even though we cannot make out its details. The earliest time for which we can gain a clear view of things using light of any wavelength is when the universe was 1100 times smaller than today, at a then red-hot temperature of about 3000 K, and was about half a million years old. At that time the primordial gas of matter changed from opaque plasma, like the gas at the surface of the sun, to clear neutral gas, transparent like Earth's atmosphere but made almost entirely of hydrogen and helium. When we map the cosmic background, we are collecting photons that have been traveling unimpeded toward our telescope since this time. This is the farthest we will ever see directly to the Big Bang, unless we someday invent telescopes that are sensitive to penetrating neutrinos or gravitational waves instead of light.

The cosmic background radiation spectrum does not "remember" anything earlier than the first few weeks, except for the total amount of energy. This information is preserved in one number which stays the same as the universe expands, the total number of photons divided by the total number of neutrons and protons. It is a very large number—more than a billion. If you count particles in the universe, nearly all of them are particles of light and not of matter.[6] In the Big Bang model, this number has stayed almost the same since the matter itself was created. It is a large number because matter when it was created was only a trace contaminant born of the light.

------

[6]If you count mass instead of numbers, the atoms dominate the microwave background by about 1000 to 1 today, but in the past the radiation dominated the mass as well. Note that nearly all the photons in the universe even today come from the Big Bang. All the starlight from all the stars that ever lived, though it contributes a similar total energy to the cosmic background radiation, has only a thousandth as many photons.

6 Primordial Matter

n addition to the cosmic background radiation, the Big Bang left behind another important relic: all of the matter in the universe today. The composition of this matter contains clues to its origin.

Although the radiation spectrum has "forgotten" details of the first few weeks except for the relative amounts of matter and radiation, the Big Bang has left measurable imprints of much earlier events. Some aspects of the nuclear composition of matter— what elements the matter is made of—provide a detailed fossil record of events when the universe was only one second old and its temperature was over ten billion degrees, a thousand times hotter than the center of the sun.

The abundances of light elements are one such fossil, especially those of the simplest elements: hydrogen with one proton, helium with two, and lithium with three. It is a profound fact that the universe is made almost entirely of hydrogen and helium, and this is because of what happened early in the Big Bang.

Each nucleus can come in different stable isotopes, depending on the number of neutrons. Hydrogen, for example, can occur with just a proton or, as deuterium, with one neutron attached; helium usually occurs with two protons and two neutrons but can occur in a light stable form, ^3He, with only one neutron. Many heavier nuclei are also stable, the stable forms always having a number of neutrons close to the number of protons. The strong interactions allow neutrons to hold the electrically repulsive protons together, and the presence of the protons keeps the neutrons from decaying by weak interactions.

Cosmic nuclear evolution, which since the Big Bang has occurred mostly in stellar interiors, steadily extracts energy from nuclear interactions and drives nuclear abundances slowly toward the lowest-energy nucleus, iron-56, with 26 protons and 30 neutrons.[1] The actual mix of various nuclei—the abundances of the different isotopes of all the elements—is a fossil relic of cosmic nuclear history. The heavy elements are records of the nuclear activity of stars over billions of years. The universe is still made largely of hydrogen

------

[1]Nuclei more massive than this tend to fly apart, and indeed, very massive ones cannot stay together for even a brief period. Heavier nuclei do however form (even unstable ones), in natural cataclysms such as supernova explosions.

and helium, however, because those are the elements produced in a hot Big Bang; the model also correctly predicts the small admixture of deuterium and lithium. All of these abundances record information about the amount of matter and light in the early universe.

ORIGIN OF HYDROGEN AND HELIUM

The smallest, most elemental constituents of the universe date back to the earliest, simplest time. Quarks, electrons, and other elementary particles originated in dimly understood processes during the first microsecond. The gathering together of these particles into neutrons and protons, and then into stable nuclei of hydrogen, deuterium, helium, and lithium, occurred during the first seconds and minutes and is something we can understand thoroughly with known physics. We can, moreover, check our theories by performing cosmic abundance measurements for locations that have been little altered by nuclear processing in stars.

At high temperature, particles have so much energy that when they collide, they are smashed to pieces. Atoms are dislodged from molecules at temperatures of a few hundred degrees (this happens when you cook something), and electrons are dislodged from atoms at temperatures of a few thousand degrees. At higher temperatures even the nuclei of atoms are split open. At very high temperatures, the components of normal atomic nuclei, neutrons and protons, are constantly changing back and forth into each other, "cooked" by weak interactions with plentiful and energetic electrons and neutrinos.[2] The total numbers of neutrons and protons do not change, because matter is not created or destroyed at this temperature. Neutrons are slightly heavier, so there are fewer of them than protons; the matter prefers the lower-energy proton state. The cooler it gets, the more pronounced this difference becomes. This intertransmutation of the two types of nucleons stops as the universe cools below about ten billion degrees, leaving about seven times as many protons as neutrons. (See Figure 15.)

This accounts for the main fact about the composition of the normal matter of the universe: It is mostly hydrogen. Heavier elements need to have approximately equal numbers of neutrons and protons, because nuclei with too large a fraction of either are unstable. When the protons and neutrons eventually cool down enough to stick together into nuclei, which happens when the universe is few minutes old and about one billion degrees, they play a game of musical chairs. Each neutron finds a partner, but six of every seven protons (six of every eight, or three-quarters of the nucleons) cannot find partners and so remain hydrogen nuclei forever—at least until they find their way into stars. (Because hydrogen is the principal fuel of stars, this is also the basic reason

---

[2]The neutrinos are still around today and are almost as numerous as photons, but they now interact so weakly that we can't detect them directly. They may still be important—if they have even a small rest mass, they could contain most of the mass of the universe!

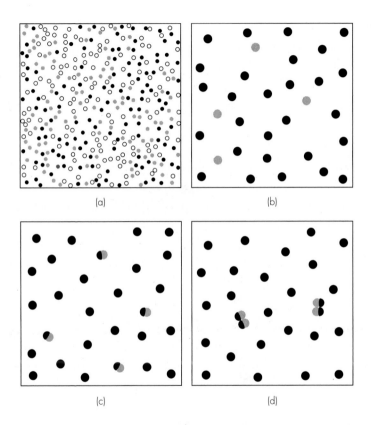

FIGURE 15 Dance of the nucleons. The baryons of the universe start life (a) as various flavors and colors of free quarks and then condense into neutrons (gray dots) and protons (black dots) as the universe cools (b). Because the protons outnumber the neutrons, most of them are left standing alone when the nucleons pair off into bound nuclei (c) and (d), leaving a universe made mostly of hydrogen and about one-quarter helium (d).

why starlight and sunlight exist in the universe today: Stars shine because hydrogen is very slowly being transformed into helium, converting some of those protons into neutrons by weak interactions.) Thus three-quarters of matter ends up as hydrogen, leaving one-quarter of matter that is roughly neutron-proton balanced to make other elements.

Only a small fraction of this quarter of nucleons—one part in ten thousand or so, depending on the density—remain as just one neutron bound to one proton: a deuterium nucleus, or deuteron. Practically all the deuterons quickly find another deuteron and bind together to form a helium nucleus. This happens to one proton and one neutron out of eight nucleons, so the Big Bang model predicts that one-quarter of the universe by mass will be made of helium and three-quarters of hydrogen, with tiny amounts of anything else. This simple prediction is remarkably consistent with what is seen. The prevalence of these light elements is direct evidence of a very hot, radiation-dominated initial state, independent confirmation that the background radiation we see is indeed from the early universe.

LIGHT ELEMENTS AND BARYON DENSITY

In these early stages, when its mass is dominated by radiation, the basic behavior of matter and radiation is determined by "just physics." The behavior of the expansion is determined by just the mass density of radiation, which in turn is determined by just the temperature, so there is only one, unique relationship between expansion rate (or age) and temperature. There is no freedom in the simplest model to adjust anything to make the temperature behave differently with time.

Even the details of nuclear composition depend on only one number or adjustable model parameter. That parameter is the ratio of the number of nucleons of matter (neutrons plus protons today), or baryons, to the number of photons. This parameter is a dial to set, which determines the mix of matter and radiation. We give this number a special symbol, η (the Greek letter eta):

$$\eta = \frac{\text{number of nucleons}}{\text{number of photons}}$$

The smaller this number, the less matter there is relative to radiation. We estimate that η is very small, just a few baryons per ten billion photons; the vast dominance of radiation over matter is why we call our model the *hot* Big Bang.

But what is η exactly? Precise calculations give exact predictions for the fraction of each light element—hydrogen, deuterium, helium, and lithium-7—that we should find from the early universe for any given value of η. We can test the Big Bang model precisely by seeing whether there is any value of η for which the predicted abundances of all the light elements simultaneously match observations, not just of helium, but of deuterium and lithium as well.

The exact value of η also tells us how many baryons are in the universe today. By measuring the cosmic background radiation, we measure directly the number of photons in the universe that are today left over from the Big Bang; there are now 411 ± 2 primordial photons per cubic centimeter of space. The value of η is almost conserved as the universe expands (we know by the thermal spectrum that not much energy has been added to the primordial radiation), so the average density of baryons today is about $411 \times \eta$ per cubic centimeter, or something like an atom or two for every ten cubic meters. The measurement of primordial light-element abundances thus not only tests the Big Bang model but also provides an estimate of η and hence of the total amount of matter. The estimate is remarkably close to the sum of all the known baryonic constituents of the universe, in the form of gas and stars in and between galaxies. (See Figure 16.)

MEASURING PRIMORDIAL ABUNDANCES

It is not obvious from personal experience that the universe is mostly hydrogen, even less so that a quarter of it is helium, an element not known to exist until the last century. We live on a planet

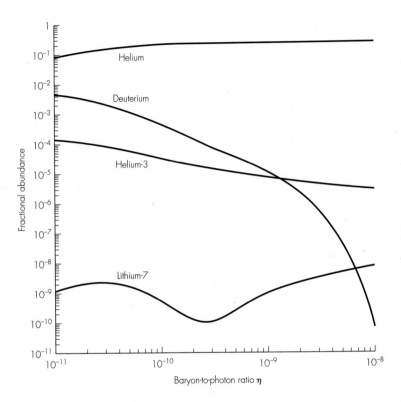

FIGURE 16 Predictions of the Big Bang theory for the abundances of light elements. The amount of each element depends on the total amount of matter η. Current data on abundances, as well as direct estimates of baryon density, indicate that there are 1 to 10 baryons per ten billion photons, near the middle of the range here. Note the very large range of abundances, from helium at 25 percent or so to lithium at less than 10^{-9}.

that is too warm to hold on to the lightest gases. Because light particles at a given temperature have a higher velocity than heavier ones, they can more easily escape a planet's gravitational pull; this is why Earth's hydrogen and helium long ago evaporated into space, except for that trapped in water and other molecules. Helium does not even form molecules, so it is extremely rare on Earth. Indeed, it exists here at all only because it is produced by radioactive decay of heavier elements.

But elsewhere in the universe, hydrogen and helium are ubiquitous. This is true in the sun, in the outer planets of the solar system, and everywhere else between the stars. Because stars convert hydrogen into helium, the amount of primordial helium is best estimated by looking at gas in nearby galaxies that appear not to have had a lot of previous star formation. These galaxies are much poorer than our own in elements made only in stars, such as iron and carbon, so they should also be poorer in any helium added by these stars to the helium derived from the Big Bang. The detailed colors of light emitted by this gas show how many electrons are recombining into hydrogen and helium atoms, yielding an estimate for the relative amounts of hydrogen and helium left over from the Big Bang. The helium fraction by mass is observed to be about 23 percent or 24 percent, close to what the Big Bang predicts. As η changes from 1 to 10 baryons per ten billion photons, the prediction for the helium fraction changes from about 22 percent to 25 percent.

Because the prediction is rather insensitive to η, the relative amounts of hydrogen and helium are excellent confirmation of the Big Bang model; the model made a clean prediction that came out right. The prediction depends hardly at all on the input parameter, which is why the helium abundance makes a good generic test of the model itself. However, using helium is not the best way to estimate η. The abundance of deuterium, the heavy isotope of hydrogen, is much more sensitive to η, so it is a more discerning probe of baryon density. Moreover, the Big Bang is probably the only important source of deuterium in the universe, so unlike helium, if any deuteurium is found, we can sensibly assume it comes directly from the Big Bang and has never passed through a star.[3]

The Big Bang leaves deuterium behind because things happen too fast for thorough nuclear burning to take place; there is time only for partial cooking. (Nucleosynthesis in the Big Bang is over in a few minutes; in places like stars, where other elements are made, nuclear burning lasts for millions or billions of years.) This allows the survival of deuterium, a fragile partial-burning product that in stars cooks all the way to much more stable nuclei such as

---

[3]This is true for all deuterium, even for the one part in ten thousand of ordinary seawater that contains a deuterium atom in place of a hydrogen. One atom in ten thousand may not sound like a lot, but this ratio makes deuterium more commonplace than carbon and iron in the universe as a whole. Deuterium was made only in the Big Bang, and the creation of elements heavier than lithium happens only in the interiors of stars where deuterium is destroyed.

helium. One can think of deuterium as a kind of partially spent fuel like charcoal, which is left over only because there is not time for all of it to burn completely to ash before the fire has cooled off. This is why more of it is left over if the Big Bang matter density is lower—the less matter, the less efficient is nuclear burning.

Measuring the primordial output of deuterium is difficult because we live in a polluted, dissipated, middle-aged galaxy that has done a lot of chemical processing over its ten-billion-year history. Deuterium can be measured in the interstellar medium of our galaxy, in clouds of hydrogen gas between the stars. However, it is very easily destroyed in stars, and because much of the gas in our galaxy has been in and out of stars many times (we don't know how many), looking at gas in our galaxy today might not be the best way to measure the primordial deuterium abundance. It would be much better if we could get hold of some really pristine primordial material.

Although we cannot bring such material into the laboratory, we *can* look at its composition. The most luminous objects in the universe, bright quasars, are so far away that the light we see now left them when the universe was less than a quarter of its present size and perhaps only a tenth of its present age. On its way to us, the light from these quasars passed through ancient pristine clouds of gas at an early time when they had not yet fallen into galaxies and had not been much affected by stars. The composition of the clouds left traces in the spectra of the quasar light, which typically display signs of relatively underabundant heavy elements and relatively overabundant deuterium, as we expect for primordial gas.[4]

The currently estimated value (about one part deuterium in tens of thousands of parts hydrogen) fits well with the predictions of the Big Bang model for the value of η, several baryons per ten billion photons, which also fits estimates of the primordial helium abundance in nearby metal-poor galaxies. Furthermore, this value is consistent with the lithium abundances in the atmospheres of nearly all the oldest metal-poor stars, a few parts in ten billion, which is thought to reflect closely the primordial lithium abundance. In other words, the Big Bang model provides a consistent prediction for all of these abundances for this value of η: It explains what primordial matter is made of.

THE COSMIC BARYON DENSITY

This consistency means that we probably understand at least some of the things that happened one second after the beginning of the universe. It also suggests that the history of matter at great distances is like that nearby, as expected in the simplest possible model of the

---

[4]By examining many such clouds in widely separated places, we can also test the prediction of cosmic uniformity: The universe should be made of the same stuff everywhere.

universe. And from the observed number of photons, our estimate of η translates into an estimate for the average density of baryons: one or two atoms for every ten cubic meters of space. This estimate agrees pretty well with the number of baryons we actually see in the universe today—all of the matter in all of the gas, stars, planets, and dust we know of, which we know to be made of normal baryonic material.

Light-element abundances are consistent with (though they do not conclusively require) a larger density of baryonic matter than we account for in known baryonic things. Such a shortfall would not be surprising, because several forms of matter that we know exist are hard to observe directly. Examples include diffuse ionized gas between galaxies, and various forms of "compact objects": cold, Jupiter-like gasballs, planets too small to burn as stars, and dark remnants of dead stars, such as black holes. What *is* surprising is that the total amount of baryonic matter counted from Big Bang nuclear composition is still not enough to account for all the mass in the universe.[5] For this reason, we think that most cosmic material must be in some new, "nonbaryonic" form.

---

[5] This statement is made in the context of the apparently successful Standard Big Bang Nucleosynthesis Model, which assumes that the matter is perfectly smooth on scales larger than single particles. By modifying the model in particular ways—for example, by allowing departures from perfect uniformity—we can reconcile larger densities of baryons with abundances, and this may allow enough baryons in the universe to account for all the observed mass. From a certain point

Dark Matter

The mean density of normal material—the average number of atoms per cubic meter of space—necessary for the Big Bang to agree with the observed abundances of light elements is enough to account for known stars and gas, but not enough to explain the total amount of mass that we know, from its gravity, must be there. If the simple Big Bang is right, then most of the mass of the universe must be some completely new form of stuff different from any known atomic substance. It is called nonbaryonic dark matter because its mass does not consist mainly of the baryons (neutrons and protons) that make up all known elements, but of some other, perhaps entirely new particle or elementary structure.

How do we know the mass is there? We can measure the mass of something by measuring its gravitational attraction—that is, by seeing how fast something falls toward it. We can, for example, measure the mass of Earth or of the sun by measuring the speed of satellites or planets in orbit around them, because things in orbit are always falling: The bigger the mass, the bigger the gravitational force, and the faster something falls, the faster it moves in its orbit.

---

of view, this possibility appears natural: Baryons today are highly nonuniform, and there are early events, such as the quark-hadron phase transition, that can introduce nonuniformities before nucleosynthesis. These ideas have not been popular, however, because they add many unconstrained parameters to the model and because most current theories about earlier events predict that baryons should indeed be uniform.

We can also measure the masses of galaxies by measuring the speeds of stars orbiting within them, because they are in orbits just as planets in the solar system are. (They are moving about ten times faster than planets, hundreds of miles every second, but galaxies are so big that even at this speed, a full orbit—a single "galactic year"—takes a few hundred million Earth years). The speed of stars and the size of their orbits tell us the mass of the galaxy. But when we weigh galaxies this way, we find that that they are about ten times heavier than the mass of all the stars within them can account for. Without the gravity of this extra mass, the stars at their high speeds would simply fly away, and the galaxies would fly apart. We also know that the extra mass is not all concentrated at the center, as it is in the solar system, but rather is distributed, as some of the stars are, in a large swarm (or "halo") of particles extending throughout and beyond the bright stellar part of the galaxies. We know this because, unlike those in the solar system, orbital speeds in the galaxies do not decrease in the outer parts—stars in the outer parts of galaxies are moving as fast as those near the center.[6]

Dark matter shows its gravitational presence in other ways. Gravity bends light rays, so a concentration of dark matter can be detected and measured by its influence on the light of objects behind it. This has been seen within our galaxy, as light from background stars is magnified by the focusing effect of dark objects in

---

[6] They have farther to go, so they typically have longer orbital periods.

the halo.[7] It is also detected very far away—for example, in the distortion of images of distant galaxies. (See Figure 17).

The cosmic nucleosynthesis constraints have stimulated many ideas for possible nonbaryonic forms of dark matter. For example,

FIGURE 17   Gravitational lensing by dark matter. This Hubble Space Telescope (HST) image shows a cluster of galaxies in the foreground. The mass of the cluster, which is almost all dark matter of some sort, has bent the light of the background galaxies, stretching the images into long thin arcs, as though they were seen through wavy glass.

------

[7]As a dark object passes in front of the star, its gravity acts as a lens, making the star brighten and then fade again as it moves away. These observations reveal that some of the dark matter in our own galaxy's halo is in quite large objects, probably much larger than Jupiter. This component of the dark matter is likely to be baryonic objects; they are dubbed massive compact halo objects, or MACHOs.

the Big Bang model predicts that the early universe produced almost as many neutrinos as photons. At very early times, these neutrinos were coupled to the photons, and although they no longer interact significantly with the photons, they are still there. But they have thinned out similarly and now have nearly the same average number density as the primordial photons. Neutrinos are so numerous (more than a billion times more common than protons) that they would constitute significant cosmic dark matter if they had even a few billionths of a proton's rest mass. Neutrinos are the only nonbaryonic dark-matter particle candidate *that is known to exist* (although their masses and hence their cosmic mass density are still not known.)[8]

It would be even more interesting if the early universe produced an abundance of some other kind of leftover particle that we have never detected in the laboratory. This is a natural speculation, given that the early universe achieved far higher temperatures than laboratories ever can and sustained those temperatures for a much longer time. For example, the universe sustained a temperature of 10^{13} K, high enough to break up baryons into quarks, for a full mi-

---

[8]Cosmic neutrinos have been detected, both from a nearby supernova in 1987 and from the sun, although the latter are not so plentiful as expected. A number of laboratory observations suggest that some neutrino types have enough mass to be cosmologically important, and there are experiments underway that could prove it.

crosecond—a billion billion times longer than the isolated reactions in accelerators last at the same energy.

If nature had stable particles that interacted only very weakly, they would be hard to produce and hard to see now, but they could still have been produced in the early universe because of the relatively long time available to make them. If they had rest mass, they might contribute enough gravitational attraction to hold galaxies together. These particles would be whizzing around us all the time at velocities of hundreds of miles per second, but they would simply pass through everything with almost no interaction. If we were lucky, a few particles of the primeval dark matter might be detected in special laboratory experiments designed to find a few extremely rare events where they did interact. These experiments could reveal new attributes of matter and possibly even a type of matter never before known—a type, moreover, that permeates everything in the galaxy and dominates the mass of the Universe.[9] The imagined candidates for cosmic dark matter extend beyond new particles; there may be configurations of force fields, miniature black holes, and stable quark nuggets, many of which have identifiable observational "signatures" but none of which has yet been found.

------

[9]This is different from discovering a new chemical element made of normal baryons. It is even different from discovering a new elementary particle such as a new quark, which is, after all, just another baryon that inhabits only rare and extreme environments.

HEAVIER ELEMENTS

The early universe manufactured almost nothing heavier than helium. A tiny portion—less than one nucleus in a billion—became lithium, with three protons, and everything else was much rarer still. Compared to the centers of stars, the Big Bang did not process nuclei very thoroughly—it quit even before it processed all the hydrogen to helium. This is because the matter of the early universe was much less dense, at a given temperature, than the centers of stars (because η is such a small number), so reactions could take place only at very high temperature. In addition, temperatures high enough for nuclear burning last only for a very short time—a few minutes instead of the millions or billions of years available in stars. For this reason, all of the other elements, including critical ones such as carbon, are made only after the Big Bang, by stars.

The universe is steadily becoming more polluted (or enriched) by successive generations of stars. Every time a star forms from interstellar gas, it burns more hydrogen nuclei to create helium and (if it is massive enough) heavier elements, then later re-ejects some of this material into the interstellar medium. Over time, as fresh hydrogen is used up, more and more baryons end up in heavy nuclei.

The oldest stars (for example, the ones used to measure the age of the universe) tend to be the least enriched. Some have less than a thousandth of the solar enrichment, and those with a tenth of the solar enrichment are fairly common. Material in the center of our Galaxy, into which enriched gas is continuously falling, has a higher concentration of heavy elements; whereas the Sun has about 2 percent of its mass in heavy elements, stars in the Galactic center can have more than twice this amount.

Although heavy elements constitute a relatively small portion of the mass of baryons, they are significant in determining the appearance and chemistry of the atomic and molecular gas and dust of the interstellar medium. Some of the enriched material ends up being collected into asteroids, moons, and planets such as our own. Because the same material is everywhere, there is every reason to believe that solar systems, including planets like Earth, are everywhere. Large planets around other stars have been detected by their gravitational influence on the stellar motion, and even if planets as small and light as Earth cannot yet be detected directly around other stars, they are probably there—at least the building materials are there.

7 The Formation of Structure

THE BREAKUP OF THE EXPANSION BY GRAVITY

The matter and radiation of the universe are not so uniformly distributed now as they were earlier. The matter is gathered into galaxies, and the galaxies cluster on still larger scales, although on the largest scales the distribution appears to be approximately uniform. Even the cosmic background radiation displays tiny imperfections—very slight hot and cold spots that probably derive from the same source as the galaxies, although they are larger than structures studied in current galaxy catalogs. The main features of cosmic structure can be explained by adding small variations in density or expansion rate to the Big Bang model and letting gravity (including that of dark matter) do the rest of the job of gathering material together.[1]

---

[1]At a deeper level, the presence of the fluctuations itself requires a physical explanation. We don't know why there are fluctuations, but explore some ideas in the next chapter. We have already seen that they exist (Figure 12).

117

Maps of the positions and velocities of galaxies in the universe show a continuous gradation from stable, nonexpanding systems on small scales to uniform expansion on the largest scale. (See Figure 18.) The "large-scale structure" that appears in the galaxy maps occupies a transition zone between the small-scale lumpiness and the large-scale smoothness of the original Big Bang. The frothy appearance of the galaxy distribution—voids, sheets, and filaments made up of thousands of galaxies—is a sign of uniform expanding motion breaking up and reversing itself randomly in different places, first in some directions and then in others, like the formation and growth of waves that eventually break into whitecaps on a choppy, wind-driven sea. This process has been proceeding for most of cosmic history; the galaxies themselves are remnants of the same process at an earlier stage.

The main force responsible for these aggregations is gravity. If a piece of the universe is expanding a bit more slowly than average, or if its density is a little higher than average, then gravity will

FIGURE 18 A map of part of the local universe, showing the positions of about 24,000 galaxies within thin pizza-slice-shaped volumes represented as dots. We are at the apex of the slices. The survey depth is a few billion light-years; the expansion velocity at the edge is 60,000 kilometers per second. There are large, conspicuous bubbles, clusters and sheets in the galaxy distribution, the frothy large-scale structure caused by gravitational instability. Above a certain scale there are no larger-scale bubbles or clusters; as with wallpaper, there is a pattern size. On larger scales every place is much like every place else; although the galaxies thin out toward the edge due to survey incompleteness, different slices look very similar.

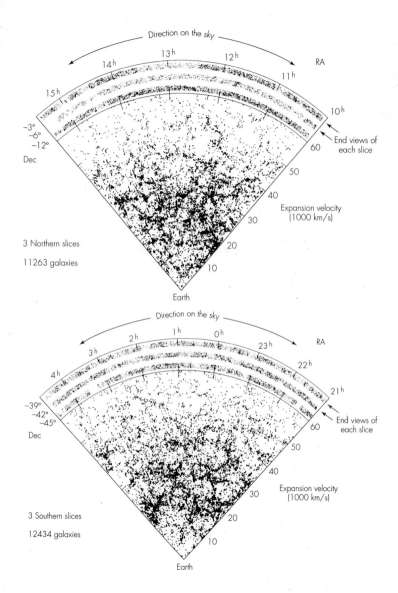

eventually make it stop expanding, turn around, and collapse. This has apparently happened repeatedly, first on small scales, causing under some conditions the formation of galaxies, and later causing their clustering into the present large-scale structure. We do not yet know *why* structure formed, because we do not know why there were fluctuations to start with, but we know a lot about *how* it formed. The physics is known, so the process can be simulated on a computer and compared with observations. (See Figures 19 and 20.)

Usually the expansion reverses in one direction before other directions, causing a collapse into pancake-like structures. Of course, if the fluctuations are random, the collapse will not occur in the same direction or at the same time everywhere, so there will be pancakes everywhere of different orientations and different stages of collapse, creating the characteristic filaments and bubbles seen both in computer simulations and the real universe. Dense concentrations of matter—first galaxies, then clusters of galaxies—form at interstices of the pancakes or bubble walls, as the matter flows into their intersections. The initially uniform gas of the Big Bang has probably done this many times, aggregating continuously into larger and larger clumps under the force of gravity. Most galaxies date from an early time, when the universe was 10 to 100 times denser than it is today. The largest-scale structures, the pancakes and filaments that meet our eye when we admire the galaxy maps, are just now forming out of the expansion; indeed, they are still expanding in some directions.

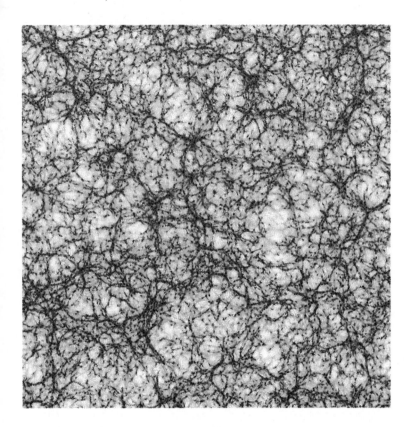

FIGURE 19 Simulation of a piece of the universe, showing the characteristic frothy appearance of an expanding universe undergoing gravitational instability. This figure shows the simulated dark-matter distribution on about the same scale as the galaxy maps in Figure 18.

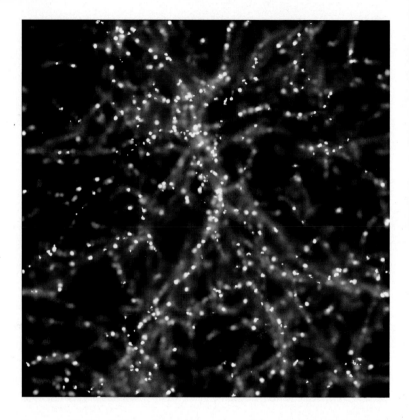

FIGURE 20 Simulation of a smaller piece of the universe, showing the same frothy appearance of gravitational instability. This simulation shows the distribution of neutral hydrogen gas on a scale of about 25 Mpc at high redshift; the bright knots are forming galaxies. This gas can be detected by quasar absorption (Figure 24).

THE HISTORY OF STARS AND GALAXIES

After gas separates out of the cosmic expansion, much of the material forms into structures called galaxies, which are quasi-stable assemblies of gas, stars, and dark matter. The wide variety of sizes, shapes, and types of galaxies reflects several competing processes in their structure and formation.

After gas stops expanding, the continuing force of gravity makes it start contracting. As it collapses, it heats up—in the same way that expansion cooled it down. Heat creates pressure that tends to resist further collapse, but at the same time, the gas is cooling by radiating energy into space. The denser pockets of gas thereby cool and continue their collapse, fragmenting along the way into smaller pieces and eventually collapsing all the way into stars. (See Figure 21.) At this stage, as we have seen, the further collapse of stars is prevented by heat, which is fueled by nuclear burning. Stellar energy also escapes and heats the surrounding gas, preventing the further overall collapse of gas into stars and stabilizing the gas of the galaxy on a larger scale.[2]

---

[2]Some of this gas is returned to the interstellar medium when stars die, but usually a remnant of material is left permanently behind, a dead cinder that joins the dark matter of a galaxy. In some cases there is no force able to withstand gravity, and collapse continues all the way to a black hole, but more often the pressure of cold electrons or nucleons stabilizes the cinder as a degenerate dwarf or a neutron star.

FIGURE 21 A nearby region of star formation out of gas in our galaxy. The light from a bright new star, off the top of the picture, is heating and evaporating the cocoons of other young stars in the dense, dark knots. The bright glowing material is mostly hydrogen gas; it is mixed with dusty dark material made mostly of heavier elements.

In gas-rich galaxies, as with stars, the tendency of matter to fall in is balanced by an outward flow of energy. The feedback of stellar energy into surrounding gas inhibits the formation of more stars, which regulates the flow of gas into stars. Because the energy is added in many ways (via radiation, explosive energy or winds, magnetically driven outflows, and so on), the detailed energy balance and stability of galaxies are not so standardized or so well understood as they are for stars. The regulation is often far from perfect, with episodes of catastrophic star formation followed by periods of relative stability. Occasionally, though less often now than in the past, galaxies crash into each other, triggering starbursts in their gas and eventually merging into one larger galaxy. These encounters also tend to eject galactic matter back into intergalactic space.

Gas in isolated galaxies tends to fall as far as it can until the centrifugal force of the spin balances gravitational attraction, leading to a spinning-disk galaxy. Without star formation, the disk of gas would get very thin and cold, like Saturn's rings of icebergs. Disk galaxies, however, tend to settle into self-regulated states where the remaining gas dribbles into stars at a rate controlled by the feedback of energy from the stars. These galaxies often take on a spiral pattern created by the differential motion of the spinning disk, so they are also called spiral galaxies (see Figure 22).

The overall tendency is for more and more gas to be locked up in stars or to be ejected to such high temperature and low density that it cannot collapse and cool. Eventually, a galaxy uses up

FIGURE 22 Part of M100, a typical spiral galaxy similar to our own. The bright knots are regions like that in figure 21, lit up by the formation of new stars out of the residual gas in the disk. The spiral pattern arises because the material inside orbits more quickly than that outside.

nearly all of its gas, and new star formation ceases. A galaxy with little gas left over becomes just a swarm of stars, each one flying in a free orbit under the influence of the gravity of all the others. In the same way that the moon is always falling toward Earth but never hits it, these stars fall continuously but, as a whole system, settle into a stable collection of collisionless particles. Galaxies like this (called elliptical galaxies) tend to be made of fairly old stars that formed from gas a long time ago and are now gradually dying off.

Our galaxy is a fairly ordinary spiral galaxy. Like most of the hundred billion or so others that can be seen with the Hubble Space Telescope, it coalesced out of the expanding universe about ten billion years ago. The gas collapsed until it was stopped by its rotation, which forced it into a flat disk shape. It continued to cool, and gravity continued to act, so that the gas disk itself began to break up still further into dense molecular clouds and, within these, into stars. This is still happening in the galaxy today, and we live close to some regions of active star formation, such as the Orion cloud and the Eagle nebula (Figure 21). However most of the original gas is now used up—locked inside stars and the remnants of dead stars. As the galaxy ages and many of the stars die, some of the material they have processed is thrown back into the interstellar medium; this is the source of the many elements (every nucleus heavier than lithium) that were not produced in the Big Bang. Stellar ejecta also steadily add more helium to the mix.

About 4.5 billion years ago (that is, about half the age of the galaxy ago), our own sun formed, together with its solar system, out of gas with considerable enrichment—about 2 percent by mass—in heavier elements. These elements contributed most of the solid particles that accumulated into rocky planets like ours. In the formation of a star, rotation forces the gas into disks like miniature galaxies, which eventually become planetary systems as the material in them collects into planets. Because of the heat close to the main star, all that is left is the stuff that is heavy and hard to boil away; this is why Earth has almost no helium and has hydrogen only in molecular combination with heavier atoms. More distant and massive "gas giant" planets, such as Jupiter, Saturn, Uranus, and Neptune, have hydrogen-and helium-rich composition like that of the sun.

We are now able to see galaxies at distances so great that we are looking back to a time long before the solar system formed. Indeed, we will never see galaxies much farther away, because we are already seeing back in time to the era of galaxy formation. The Hubble Space Telescope sees objects a trillion times fainter than a bright star—so faint that the count of observable galaxies is now a few million galaxies per square degree of arc, or about 100 billion over the entire sky. The deepest images (Figure 23) record the

FIGURE 23 The deepest image from the Hubble Space Telescope, about ten days of exposure in one direction. The most distant galaxies seen here emitted their light when the universe was less than one-sixth of its present size (that is, wavelengths stretched by the expansion by a factor of six) and 10 percent of its present age (that is, about 90 percent of the way back

in time to the beginning). The few thousand faintest galaxies in this field represent the earliest galaxies to form in this direction; they have the color and detailed spectrum expected of new galaxies of mostly young stars, in the right numbers to account for the number of galaxies in the universe today. This is a very small patch of sky, but every direction in the sky at this depth looks almost the same; if you could image the entire sky, there would be about 100 billion galaxies. The cosmic background radiation is coming from behind these galaxies.

shapes and colors of galaxies in the early stages of their lives, when most of the stars in the universe today were forming from primordial gas. The coagulating of the universe can therefore be seen directly and is consistent with the idea that galaxies and clusters assembled by hierarchical gravitational clustering out of a uniform expansion.

THE BRIGHTEST THINGS

The centers of early galaxies tend to spawn extremely bright sources of light called quasars. These are by far the largest and brightest sources of energy in the universe.[3] Some are ten thousand times brighter than an entire galaxy.

This cosmic spectacle is powered by the liberation of gravitational energy. We are familiar with this physical effect; when you

---

[3]This is true if one counts the total amount of energy, but curiously, there are sources of energy called gamma ray bursts that for a few seconds are much brighter than quasars. Gamma ray bursts occur very rarely (once per million years in an entire galaxy), probably from the belching after a stellar mass black hole eats a large meal—and indeed in this respect they resemble miniature quasars. Their gamma ray emission was first detected by spy satellites in the 1970s, but it was only in 1997, with the detection of their optical light, that their enormous distance and brightness were confirmed. Whereas quasars emit up to a billion solar rest masses in a few hundred million years, gamma ray bursts emit about one solar rest mass (that is, an energy equal to the mass of the sun times the speed of light squared) in a few seconds.

ski down a hill, you are powered by the liberation of gravitational energy. If the hill were very long, you would eventually speed up to near the speed of light, and the energy of your motion would be close to your rest energy, which is given by Einstein's famous formula $E = mc^2$, where m is your mass. That is a lot of energy. If you smashed into something at this speed, you would create an explosion equivalent to the detonation of about a thousand nuclear warheads.

Of course, there is no hill on Earth long enough for you to reach the speed of light by skiing.[4] Very large amounts of gravitational energy can be extracted only from an extremely compact body such as a neutron star or black hole, where the mass is concentrated in a small volume. A black hole the size of Earth would, for example, have a mass 2000 times that of the sun; if you fell into one of these, you would be traveling close to the speed of light.

In the case of quasars, the gravitational energy comes from material swirling into a giant black hole between a million and a billion times the mass of the sun and similar in size to the orbits of planets in our solar system. The black holes are still there in the centers of many galaxies today, but the quasars are not so bright or so numerous as they were in the past because now there is less

---

[4]Note, however, that the velocities of objects falling from outer space are still high enough to be destructive, typically tens of kilometers per second. Very large things falling from the sky have caused explosions large enough to trigger mass extinctions on Earth.

material falling into them. The gravitational energy liberated per mass m approaches the rest energy of the matter according to $E = mc^2$: In bright quasars, a billion times the mass of the sun is converted to visible energy this way with an efficiency a hundred times that of nuclear technology. The energy is extracted in many forms: Infalling gas is heated by collisions and shocks, and the spinning energy of the black hole (imparted by the gravitational energy of infalling material) drives an electromagnetic dynamo that beams jets of electrons and positrons outward at close to the speed of light, creating radio beacons larger than an entire galaxy. The population of quasars creates an intense background of ultraviolet and X-ray light that reionizes the intergalactic gas throughout the universe.

THE DARKEST AND EMPTIEST PLACES

In spite of the fantastic character of the quasar energy source (or central engine, as it is called), for cosmology their most interesting application is just to serve as bright distant light sources. Quasars appear early in the formation of galaxies and are so bright that we can see them clearly and even measure their spectra with high precision. Ironically, it is by observing these brightest sources that we learn the most about the emptiest and darkest places in the universe—by looking at the detailed structure in the spectrum of the light modified by passing through rarefied intervening gas.

The light we see today travels a great distance through the universe on its way to us. Some of it is absorbed or scattered by the gas it passes through en route, leaving gaps in the spectrum of the light that reaches us. These gaps or absorption lines, which occur at wavelengths that depend on the number of atoms of each type present at each cosmic redshift along the line of sight to the quasar, give us detailed information about the distribution and composition of the gas for the entire history of the universe in that direction since the light left the quasar.

The absorption spectrum of light of quasars shining through the gas reveals particular patterns of absorption that can be identified as atoms of hydrogen, deuterium, and heavier elements. Atoms at many different redshifts lead to the recurrence of these patterns from different concentrations of gas along the line of sight. Smooth gas absorbs a continuous band of wavelengths, whereas gas collected into discrete clouds and galaxies produces discrete absorption lines at the particular redshifts where the clouds happen to lie. In these systems, the abundances are revealed by the relative absorption for each atom. In this way, the history of the condensation and enrichment of primordial gas is laid out for us in the spectra of distant quasars—in the absorption patterns of hydrogen, helium, and other elements in numerous intergalactic and protogalactic clouds along the line of sight. A high-quality quasar spectrum contains thousands of absorption lines (Figure 24).

Hydrogen, for example, absorbs light efficiently at a particular wavelength (121.6 nanometers) called Lyman-α, which corresponds to photons of the precise energy required to excite the electron of a hydrogen atom to its first excited state. Hydrogen is the most abundant element, so it is not surprising that Lyman-α absorption appears in light from a typical quasar hundreds of times, each time from a different cloud along the line of sight, recording the history of the conversion of uniform gas from the early Big Bang into the discrete galaxies we see today. Lines of other elements trace the evolution of cosmic chemistry from the pure light elements of the Big Bang to the polluted present.

Telescopes in space have recently enabled us to measure absorption by cosmic helium, whose lines were previously in the inaccessible far-ultraviolet region of the spectrum. Compared to hydrogen, helium atoms are more highly charged and therefore better able to hang on to their electrons in the cruel ionizing conditions of intergalactic space. This makes absorption by helium detectable everywhere—not just in clouds but even in the emptiest regions between the galaxies. Using helium absorption, we can map the distribution of the last traces of primordial gas that have not fallen into galaxies. Quasar spectra have thereby revealed that even the darkest, emptiest places in the universe contain detectable signs of

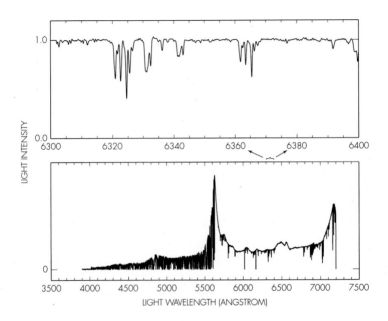

FIGURE 24 A quasar spectrum taken with the W.M. Keck telescope. The spectrum of the light when it left the quasar was quite smooth except for a few broad humps; all the fine detail was introduced as the light passed through material along the line of sight. The upper panel shows a small portion of the spectrum, revealing a repeated pattern caused by different atomic transitions in the same material. The left side of the larger portion in the lower panel shows the "Lyman-α forest" of hydrogen absorption lines, caused by hundreds of gas clouds at different redshifts. Each line is caused by a cloud that the light encounters on its way from the quasar to us. Some of the lines are suitable for measuring deuterium or heavy-element abundances.

matter. There is no place in the universe that is completely empty of gas, even though there are large regions that appear to be devoid of stars and galaxies.

CONDITIONS FOR COSMIC SELF-ORGANIZATION

When a kid makes a mess, the mess never disappears on its own. This is actually a law of nature: There is always a tendency toward randomness. Hence it seems paradoxical that structure and information could arise in a universe that was at first microscopically completely random.

How did a system that started out as simple and uniform as the Big Bang ever develop interesting complexity? Why didn't it stay totally uniform? There is no place "outside" for the information to have come from, so the universe must have become complex "on its own."

The most interesting complexity, of course, is our own selves and other living things, incredibly complex molecular systems. It seems a miracle that pure physics can make all of this happen. But in spite of much recent progress in understanding complex systems, there is no clear evidence that any new principle of nature is needed to create complexity—it "just emerges" under certain circumstances with the standard ingredients of standard physics. The debate on whether some sort of new organizing principle is required recalls the revolution precipitated by Darwin, who showed in detail how purely blind, mechanistic forces of selection could lead to the development of new species: Life is a naturally occurring property of certain complex open systems.

We now know that the physical universe began not only sterile of life or intelligence but also nearly sterile of structure or even information of any kind. *Everything* unfolded via a process of cosmic evolution from a hot light bath of sterile elementary particles. Somehow the expanding universe has provided the opportunity for the growth of structure, including the chemical and energetic opportunities we have exploited by natural selection, during the interesting events of the last four billion years on our planet, to arrive where we are.

Big Bang cosmology indeed provides the necessary critical ingredients for interesting things to happen—for the universe not to remain boring gas and blackbody radiation. The key active ingredient is the expansion itself, which allows the universe to leave thermal equilibrium and avoid complete stagnation (or "heat death"). The residue of light elements is one of many examples of this; an equilibrium universe (one that expanded much more slowly than allowed by gravitation) would be made of iron, with no fuel for stars.[5] The dark sky at night is another example of how

---

[5]Stars also create their own important departures from equilibrium. For example, stellar explosions can generate nuclei heavier than iron, such as uranium, that are unstable. The energy stored in such nuclei is later liberated during nuclear fission—in our nuclear reactors, for example.

the expansion creates opportunities for self-organization: Without a cool, dark sky to radiate excess heat into space, we would stifle in thermal equilibrium with no free energy available to do anything. Although thermodynamics texts sometimes refer to the heat death of the universe, in fact it had a heat *birth*—the closest approach to global equilibrium was near the beginning, and since then the expansion has been creating departures from equilibrium. Thermodynamic equilibrium will now never be achieved; the expansion is thinning things out too quickly.

Another essential ingredient is the growth of macroscopic structure, the assembly of matter into lumps bigger than atoms, caused by gravitational instability and by fluctuations that trigger it. Once gravitationally bound systems (such as planets, stars, or galaxies) form, an interesting property of self-gravitating systems makes them particularly good engines for generating further departures from equilibrium—they tend to heat up as they lose energy! The extreme heat liberated by material falling into black holes is just one example of this general principle, which represents a long-term trend: With time, more and more matter will be locked up in more gravitationally bound systems, providing an inexhaustible supply of free energy.

The growth of structure sets the stage for cosmic evolution to take place around stars, on planets, and elsewhere. These examples show how the cosmic expansion makes it possible for the universe to organize itself—it can grow its own complexity "from within" without outside help. Physical cosmology is now discovering the details of how a simple universe created its own complex internal organization, much as paleontologists have revealed the evolutionary path of biological systems on Earth.

8 The Beginning

COSMIC INFLATION

If we can explain why the universe is expanding, we may also understand other important facts. For example, why is it so much bigger than, say, an atom? This is not a silly question, because the laws of physics seem to allow universes of any size. It is an important question because clearly the universe must be a great deal bigger than an atom to have much interesting going on inside it and to last a long time. But we have already seen that the range of cosmic scales is not infinite. What determines its size?

The main force of nature that controls motions on a large scale is gravity. In the universe today, we see gravity only as an attractive force that holds the solar system together, as well as the galaxy. But there are conditions under which gravity can be a repulsive force, like the force pushing apart the north poles of two magnets. A magical effect of this repulsive gravity is that it can make the universe fly apart—and cause the start of the Big Bang. This effect is the central idea of a model called the "inflationary universe."

The possibility of repulsive gravity arises in Einstein's General Theory of Relativity. Einstein showed that all forms of energy are equivalent and that all of them couple to gravity. Not only does matter create gravitational forces, but so do forces. Even the forces that resist gravity, such as the pressure of matter that keeps Earth from collapsing, also contribute a certain amount of gravitational attraction. Because gravity gets stronger between particles as they get closer together and is at its strongest in the densest matter, there are conditions in which even the pressure forces, as they try to resist collapse, themselves contribute so much energy (and hence more gravity) that they cannot possibly win. These pressure forces defeat themselves: The harder they push out, the more their own gravity pushes in.

Consider where this paradoxical situation leads: The more pressure there is to resist gravity, the more gravity there is to resist. This feedback is responsible for the runaway victory of gravity in the formation of black holes, where all the matter is concentrated by gravity into a single point and its original mass is entirely consumed by the gravitational field itself.[1]

---

[1] In more moderate situations, the gravitational field consumes a small portion of a particle's mass, in a phenomenon known as the gravitational redshifting of mass. Paradoxically, a particle has slightly more mass in space than on the surface of Earth, in spite of its weightlessness in space.

In cosmic inflation, a similar situation occurs, but in the opposite direction—the runaway instability goes in the direction of making everything fly apart very fast. This can happen because pressure can actually be negative instead of positive.

Negative pressure itself is not so odd; it is just a tension, like that in a rubber band, a suspension bridge cable, or a bungee-jumping cord—a force that pulls together rather than pushes apart. The odd thing is that a tension adds a negative gravity; in all of these examples, a small amount of repulsive gravity is generated by the tension. Normally, the tension is so small compared to the density that it has no effect. But by the same paradox that makes a positive (push things apart) pressure cause attractive (pull things together) gravity, a negative (pull things together) pressure causes repulsive (push things apart) gravity, which means that when the pressure or tension is very large, instability develops: The bigger it gets, the more it wants to get bigger. The first situation, when attractive gravity runs away out of balance, leads to a black hole; the second leads to an inflationary universe.[2]

---

[2]Small amounts of repulsive gravity abound but are ordinarily overwhelmed by the attractive terms. In Einstein's theory of gravity, the formula for the acceleration of a particle on the edge of a sphere of matter of radius r is

Gravitational acceleration = $(4\pi Gr/3) \times$ (density + 3 $\times$ pressure $\div c^2$)

where c^2 is the speed of light squared and G is Newton's gravitational constant. Normally, the pressure term can be ignored until particle velocities approach the speed of light, but it is always true that positive pressure adds to the gravitational

The kind of stability or instability is determined by the type of matter present. Conditions where a negative pressure causes repulsive gravity—which requires a tension of magnitude more than $-(\frac{1}{3})$ times the density times c^2, a very large number—do not normally occur. But they could occur in the nearly empty space between galaxies. Einstein recognized this possibility, calling it a cosmological constant.

In modern terms, we say that the physical vacuum, the state with no actual particles in it, may nevertheless have nonzero gravitating energy. "Empty" space, completely evacuated of all forms of matter, may not be completely empty; when it is as empty as you can make it, there may still be energy there. In keeping with this seeming paradox, we could call it "vacuum matter." If empty space does have energy, it is associated with a negative pressure that tends to make the expansion of the universe accelerate.

An inflationary epoch at the beginning, however, which must occur before nucleosynthesis and therefore must be over in much less than a second, requires a huge negative pressure, and hence a much larger vacuum density, than is allowed today. Thus the energy of the vacuum must change from big then to small now. This possibility used to be regarded as not physically plausible. But advances in

---

effect of the density, whereas negative pressure or tension opposes it. A stretched rubber band, if it had no mass other than its tension, would exert repulsive gravity.

our understanding of the fundamental forces of physics have revealed that fields with this behavior must exist. Indeed, they lie at the heart of breaking the symmetry between the weak and electromagnetic interactions and explain why particles can have the masses they do. The vacuum is *required* by the theory to change its energy from very large at high temperature (the "false vacuum," because it is not the true lowest-energy vacuum) to much smaller at low temperature (the "true vacuum"). It is therefore not far-fetched to imagine that at the beginning of the universe, the energy of the false vacuum not only was present but was actually dominant over the density of positive-pressure material such as matter and radiation.

If this happens, then the repulsive force of gravity in even a tiny piece of space is very large. Two bits of energy with a small separation are pushed apart, and the continuing force of gravity pushes them apart with ever greater speed. A small volume of space quickly grows to enormous size in a rapid explosion. This runaway instability is the first magical property of inflation. It is why the universe is much bigger than an atom.

Curiously—and this is the second magical property of inflation—the density of material does not decrease during this vast expansion. Normal matter, of course, thins out if it expands, but inflationary vacuum matter does not. This "free lunch" effect seems to violate the principle that energy is never created or destroyed but only changes from one form to another. Here you start with a tiny speck of vacuum matter and end up with a whole universe of it.[3]

Again, the situation is a mirror image of a black hole collapse. The "free lunch" of inflation is not very different from the energy quasars get for free by dropping material into their black holes. There, matter falls into the central point and is destroyed, while its energy is transformed into "pure gravity": A black hole is made up of empty space where all the mass-energy is in the curvature of space, in the gravitational field. Indeed, it is possible to add matter carefully to a black hole in such a way that its mass is not added to the hole—the mass of a particle is separated by gravity from the particle itself. In inflation, the reverse process occurs: A large volume of energetic material (excited false vacuum, which can eventually be converted into real light and matter) is created by the action of (repulsive) gravity. Although this occurs seemingly "from nothing" (or almost nothing), the large energy is created at the expense of a large negative gravitational energy, so the final energy of the whole universe is still actually very close to zero. But it

---

[3]The Steady State Cosmology model, an expanding universe model that had its heyday in the 1950s, anticipated many of the ideas of modern inflationary cosmology, including this idea that the universe expands but the matter does not thin out. In that model, the inflation occurs today, along with continuous creation of matter from the vacuum. We now know that the Big Bang started very dense and hot, so both the origin of the expansion and the origin of matter are now relegated to exotic early epochs, but the gravitational physics is very similar in the two models.

has been transformed into a very different form of almost-zero energy from being a tiny speck of almost nothing; it is a large universe made of vacuum matter together with its gravity.

CREATION OF LIGHT AND MATTER

The weird form of repulsive-gravity "vacuum matter" that caused inflation does not make up our present universe. Its energy was long ago transformed into more mundane and familiar things. Somehow or other (there are many ideas about exactly how), the energy of the inflationary vacuum converted (or "reheated") into light, the same light that now, redshifted by a very large amount, makes up the microwave background we see today. The large universe of vacuum matter became a large universe filled with radiation energy. Radiation has a large positive pressure, so the force of gravity became attractive, ending inflation. The expansion stopped accelerating and started slowing down, and as the radiation expanded, it cooled off and thinned out.

After reheating, this model universe strongly resembles the large, expanding, radiation-dominated hot Big Bang that describes our own universe at high redshifts. It still lacks one critical ingredient, however: There is no net matter in it. Vacuum decay creates precisely neutral radiation, with exactly the same number of particles and antiparticles, baryons and antibaryons. If such material had expanded and cooled to low temperature, the particles and antiparticles would have annihilated each other, and there would be nothing left but light. This won't do, first because we know baryonic matter exists (for instance, as stars, planets, and people), and second because Big Bang element formation requires a certain ratio of baryons to photons to get the light-element abundances right.

Why does the universe have matter in it? Sometime after inflation, something happened in the universe to make baryonic matter out of pure light. Something made more quarks than antiquarks in a system where the numbers were initially the same.

To make matter out of pure light requires three ingredients, all of which are thought to have occurred in the early universe. First, there has to be a physical process that can create a quark without an antiquark. Although this is regarded as likely to occur, it must be very rare, because it implies something that has never been seen: the spontaneous creation of quarks from energy, or its reverse, the decay of baryonic matter into light.[4] Second, matter

---

[4]Because matter apparently at one time emerged from an equal mix of particles and antiparticles, the possibility of the reverse process implies that matter is unstable and will all eventually decay back into an equal mix of particles and antiparticles. Ordinary radioactive decays certainly emit light, but they always preserve the number of neutrons plus protons, or the total number of quarks that make them up. Ordinary protons and neutrons are like little spring-loaded bombs; if they could dump their quarks, they would just fly apart, releasing nearly all of their rest mass into energy. It is thought that this indeed happens, although so slowly that it has never been detected, in spite of impressive experiments conducted in an effort to observe it—experiments that tell us the lifetime of a proton exceeds 10^{31} years.

and antimatter have to behave differently; otherwise, they would always be be generated in exactly equal numbers. Fortunately, they *are* observed to behave differently, though by a very small amount. Finally, the universe has to go "out of equilibrium" to give these effects a chance to act. The expansion has to bring the particles out of their most random arrangement; otherwise, particles and antiparticles, which have the same mass, would always be equally abundant in spite of their unequal interactions. This too can happen under the right circumstances. Exactly how these ingredients are realized together in nature is not yet understood; it is a part of the Big Bang model that is still unfinished. (When this is understood, we will know the reason for the tiny value of the baryon-to-photon ratio η.) Understanding the creation of the baryon excess may eventually give us a window into the very early Big Bang and the new physics active within it, like the information from the creation of the light-element nuclei, but even earlier.

WHY THE UNIVERSE IS JUST SO OLD

Why is the universe 10 or 20 billion years old? Why not 10 or 20 trillion years? Why not 1 year? The reason why the Universe lasts *at least* 10 or 20 billion years is the same as the reason why it is much bigger than an atom; the inflationary process that made the universe large also gave it enough expanding momentum to last a long time. But why is it not even older than it is?

A clue is the age of Earth and the sun: 4.5 billion years or so, already a third of the entire age of the universe. If the Universe were much younger than it is, there would not have been time for the galaxy to collapse, for stars to make heavy elements, for the solar system to form in the galaxy, and for life to evolve. All of these things take billions, not millions, of years. Thus, given that the universe stays around for at least a few billion years, we (that is, organisms that require a sun and planet to exist and some time to evolve) could not have expected to show up much earlier than we did. At the same time, we might not expect to be around much later, because the fresh hydrogen gas that fuels the stars is running out, and what remains will be largely used up in many billions of years. After another 10 billion years, the universe will be dark and cold compared to the conditions that prevail today. We should not be surprised that we appear in the universe at the time most amenable to our being here; the universe is just so old because both chemical evolution and biological evolution lie in the critical path for the development of people.[5]

------

[5]There is no reason we know of why biological evolution should have the same timescale as stellar evolution, so it's still puzzling that we people should appear on the scene as the sun is about halfway through its main-sequence lifetime. This coincidence might have an "anthropic" explanation.

A slightly different issue is why these things take so long to happen. That is, why *does* the chemical evolution of a galaxy take billions of years? The answer lies in the physics of these systems—for example, the physics governing how stars evolve. Stars are so big, and they last so long, because the force of gravity is so weak compared to the other forces of nature. It takes a lot of atoms together before the force of gravity can be noticed at all compared to the electronic forces that hold atoms apart. Therefore, very large assemblies of atoms[6] are needed before they can be compressed and heated by gravity to the point of nuclear burning. Such a large number of atoms has enough nuclear fuel to last a very long time and uses that fuel only slowly, because it takes radiation a very long time to leak out of something as large as a star.[7] The scope of the universe in space and time (the range of scales we sketched in Chapter 2) can be traced to these large numbers of fundamental physics.[8]

---

[6]"Very large" here is given roughly by a specific number, the "Chandrasekhar mass," $(Gm_p^2/hc)^{-3/2} \approx 10^{57}$ atoms. This mass is determined by the fundamental constants: G is Newton's gravitational constant, m_p is the mass of a proton, h is Planck's constant governing quantum mechanics, and c is the speed of light. We can also write this large number as $(m_{\text{Planck}}/m_{\text{proton}})^3$.

[7]This timescale is also determined by fundamental constants. Very roughly, it is a microscopic proton-oscillation time $h/m_p c^2$ times the large number $(Gm_p^2/hc)^{-1} \approx 10^{38}$, or $(m_{\text{Planck}}/m_{\text{proton}})^2$; or $t_{\text{Planck}} \times (m_{\text{Planck}}/m_{\text{proton}})^3$.

[8]The number of different places in the observable universe is about $(m_{\text{Planck}}/m_{\text{proton}})^9$, the number of different times is about $(m_{\text{Planck}}/m_{\text{proton}})^3$, and the number of different events is the product of these.

WHAT CREATED NONUNIFORMITY?

We have seen how, if the Big Bang began with small imperfections, gravity completed the job of making galaxies. But what was the source of the initial fluctuations?

The initial kick of the Big Bang expansion came (maybe) from the repulsive gravity of a hypothetical new field, a new kind of force. One idea about the origin of cosmic fluctuations is that they were created as nonuniformities in this field during inflation itself. Such fluctuations are inevitable at some level in quantum mechanics, and in some models the quantum fluctuations can feed into fluctuations of the expansion speed.

If this idea is correct, then the universe's structure originates in its birth; the energy it took to assemble great clusters of galaxies is an effect of microscopic physics. The irony is that the largest-scale astronomical structure displays a frozen picture of enormously amplified quantum fluctuations from the smallest scales of space-time.[9] (See Figure 25.)

Another possibility is that other unknown fields are causing important effects in the universe even now. Perhaps the early infla-

---

[9]Not even the lowest-noise laboratory amplifiers can achieve this effect. Inflation can in principle act as a huge and very quiet amplifier—quieter than any yet built. For this reason, it is not possible to propose an accurate physical or laboratory metaphor for the process; coherent amplification of quantum vacuum fluctuations has never been observed.

tionary period left a very smooth universe behind; perhaps instead of fluctuations dating back to the beginning, space is filled today with invisible fields traveling very fast across intergalactic space, stirring up matter by the gravity of their invisible disturbances. If the large-scale structure is like whitecaps on stormy water, then these new fields are like the invisible wind that causes them. Such active scalar fields could be responsible for the organization of matter into structures, including the formation of galaxies and large-scale structure. In this case, cosmic structure is not really a microscopic effect; there is instead new physics that involves exotic fields operating on large scales today.

Spatial variations in scalar fields form from a process in the early universe that resembles a huge grid of falling dominoes or pencils connected by springs (Figure 26). The state of the field at every point is represented by the pencil there; the springs represent field forces that "try" to smooth things out. At high temperature, everything is flailing about wildly, and the average position of the pencils is straight up; when the system is cooled, they fall down. But if the cooling occurs suddenly, they fall down in different directions, only later lining up under the action of the springs. This "self-ordering" takes a long time on large scales, because it can propagate only at a certain speed, which is much slower than the time it takes a pencil to fall.[10]

---

[10]Special "topological" defects, such as cosmic strings or monopoles, occur if there is a cowlick that can't be combed out—a single pencil surrounded by others all facing away from it.

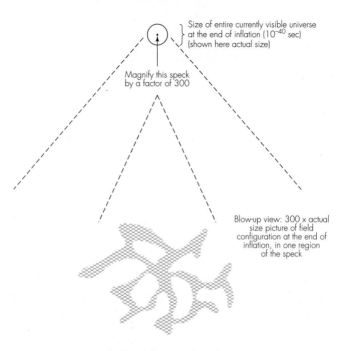

Size of entire currently visible universe at the end of inflation (10^{-40} sec) (shown here actual size)

Magnify this speck by a factor of 300

Blow-up view: 300 x actual size picture of field configuration at the end of inflation, in one region of the speck

After fifteen billion years, this leads to:

Tiny quantum fluctuations have become galaxy superclusters

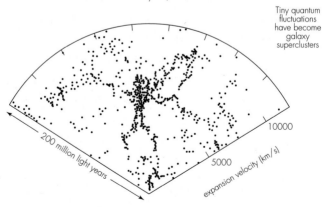

200 million light years

expansion velocity (km/s)

10000

5000

154

FIGURE 25 A cartoon showing structure formation from quantum inflationary fluctuations. The dot at the top shows the actual size, just at the end of inflation, of all the matter in the observable universe today. The age of the universe at this time was *much* less than the time it takes light to cross the dot (in fact, it is smaller than this by about the same factor by which the dot is smaller than the universe today). An enlargement (about 300×) of a small section of the universe at this time (middle) reveals a fluctuation in the field that controls the inflationary vacuum energy; this was created still earlier, by a quantum vacuum fluctuation. Because of this fluctuation, this section of the universe is expanding not precisely at the average rate, but a little more slowly, so it always falls a little behind. Eventually, it collapses to form the Coma supercluster of galaxies (bottom).

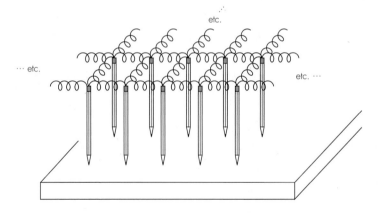

FIGURE 26 Symmetry breaking, illustrated with a table of pencils swiveling on their tips, their eraser ends connected by springs. The state of a field at each point in real space (two dimensions of the table top pictured here) is represented by the tilt of a pencil at that point. Although the table top is symmetric (all directions are the same), the upright state of the field shown here is unstable, so even if they are all standing straight up at the beginning, they end up falling down in some direction or other. If the pencils fall down quickly, they will not have time to align and will fall down in different directions, after which the springs will jiggle like a cheap mattress. This causes fluctuations in energy whose gravity creates fluctuations in ordinary matter.

Spatial variations in scalar fields store energy like the network of unequally stretched springs, and as the energy is released, it creates fluctuations by gravity. The relaxation or self-ordering of these variations is like the pencils lining up, and it leads to large-scale oscillations that create fluctuations in matter and radiation. This self-ordering and the movement of matter and energy happen on large scales at late times—hence the term *active* scalar fields.[11]

This is perhaps the widest range of uncertainty in science: We don't know whether cosmic structure is ultimately caused by something happening on scales far smaller than atoms or on scales the size of the entire universe! The view currently in favor is that the fluctuations derive from the quantum mechanics of inflation. This theory will be tested soon as new, detailed maps of the background radiation become available.

---

[11]A most peculiar possibility of this kind is cosmic strings. According to this idea, some types of cosmic strings extend like vast wires across intergalactic space and wreak havoc at close quarters. As they whip around at nearly the speed of light, the two opposite sides of the sheet swept out by a string suddenly find themselves moving toward each other with a speed of about a kilometer per second. This kind of interaction can lead to observable effects, such as sharp edges in the microwave background. It is important to bear in mind that such structures may or may not exist; if they do exist, however, they exhibit predictable behavior.

9 The Future

ur attention in this book has been focused largely on our past, but it is time to ask whether physics and Big Bang cosmology have anything to say about the future, especially about the very far future.

The expansion itself has two fairly simple global options: Either it can go on forever, or it can eventually stop expanding and then recontract, ultimately collapsing to a very high-density, hot state.

If it does recontract, it will not turn around for a long time, and there are still tens of billions of years until the "Big Crunch." The final state will be very different from the initial one because of the growth of structure during cosmic evolution. The final state will not be smooth and simple like the initial one but will be bumpy and chaotic, full of turbulence and complexity.

The Crunch option seems less likely today than it once did. As we have seen, comparison of cosmic age and expansion rate implies little slowing, so the force of gravity is probably too weak to reverse the expansion. In this case, things will cool down over time and thin out. In this scenario the remaining fresh gas is gradually consumed, removing the raw material for new stars; the universe gradually becomes darker, more enriched in heavy elements; new stars stop forming; and baryons are locked in cold remnants and black holes. The universe cools off into a "cosmic winter."

It is also possible that in the very distant future, when we can see very far away, we will discover that the simple Big Bang picture is only a good approximate description of our local patch of the universe and that things look quite different on very large scales—perhaps highly chaotic, or asymmetric, or rotating. That we can as yet see only part of the universe means that guesses about its very long-term future will remain just that. Nobody can be sure what is about to drift into view over the particle horizon.

THE FUTURE WITHIN THE UNIVERSE

Given that the universe seems certain to be around for at least several more tens of billions of years, the fate of the average motion of matter is not the most interesting thing about the immediate future. What we really care about is the future of life, complexity, intelligence, culture. The vision of a cosmic winter sounds bleak, but we should remember that there is plenty of time for intelligent beings to adapt to the new conditions. We can "learn," in the evolutionary sense, to be comfortable. No matter how cold it gets, there is no physical reason for thought or activity to stop. Things may slow down, and little that we would recognize of our present form may persist, but there could be plenty of time and plenty of gravitational energy available for an infinite number of thoughts and perceptions.

The bleakest version of the far future unfolds if matter decays—that is, if protons and neutrons are unstable and fly apart into lighter particles—because then there are eventually too few atoms to construct intelligent beings. Even in this case, however, information and perhaps even intelligence can survive as long as there are some stable particles left to interact with light. For example, electrons and positrons are thought to be absolutely stable, and if so, life will find a way to re-engineer itself by adapting its activities into a new medium made just of those particles.[1]

Even though in some models we can look forward to an infinite number of thoughts in the future, there are still choices to make; it is not true that all thoughts will eventually be thought. Some infinities are bigger than others, and an infinite universe that lasts an infinite time is still not big or long-lived enough to realize

---

[1]In models with nonzero cosmological constant, however, even stable matter particles are flung apart from each other so fast that they can't communicate, eventually freezing into eternal solipsism.

all of its possibilities. For this, we need an infinite number of infinite universes.

Comparing infinities requires some care. There is an infinite number of positive integers—the sequence $1, 2, 3, 4, \ldots$ goes on forever. But there is a bigger infinity than this, even if we stick just to numbers: the infinite number of real numbers, comprising all infinite, nonrepeating decimals, which include some numbers with special names (such as $\pi = 3.1441592654\ldots$, $e = 2.718281828\ldots$, and $\sqrt{2} = 1.414213562\ldots$) but also all the other possible sequences. The infinite number of sequences is much bigger than the infinite number of integers.

To prove that the infinity of real numbers is larger than that of integers, consider a mapping between the two sets (called in this context a Cantor sieve). If two infinities are equivalent, then it must be possible to make a one-to-one mapping between them. For example, the positive even numbers are an infinity equivalent to the positive integers, because we can write a mapping where each even number is assigned to an integer:

1. 2
2. 4
3. 6
⋮   ⋮

and so on.

If the real numbers were equivalent to the integers, we could find a similar mapping, such as

1. $\pi = 3.1441592654\ldots$
2. $e = 2.718281828\ldots$
3. $\sqrt{2} = 1.414213562\ldots$

⋮ ⋮

and so on, where each integer $1, 2, 3, \ldots$ labels a certain real number represented by an infinite decimal.

Now we can show that there is no such mapping where every real number is assigned to an integer. For any attempted system, we can always find a real number that is not included, by choosing the first digit of the decimal not to equal the first digit of the first number (in this example, the first digit of π, or 3), the second digit not to equal the second digit of the second number (in this case, the second digit of e, or 7), and so on. The resulting number is not included in the table because it always differs by at least one digit from any entry. Therefore, the infinity of real numbers (or of all possible infinite sequences of integers) is larger than the infinity of the integers themselves.

A similar argument can be made for cosmic information. The states of atoms, molecules, and other particles can be described by numbers; a particular configuration can be represented by a sequence of numbers; and a sequence of activity in a finite volume of space, but stretching to infinity in time, can be represented by an infinite sequence of numbers, like a nonrepeating decimal. The volume of the universe in space and time may also be infinite, but it is not so infinite as this—it can never grow large enough, fast enough to contain all the combinatorial possibilities. To explore all the possibilities, even with a fixed set of physical laws, would require an infinite number of universes. Similar arguments indicate that far from being an autonomous "robotic" universe, the world is both unique and unpredictable—the simple initial state of the universe can't possibly contain enough information to predetermine its complex future evolution.

THE ANTHROPIC UNIVERSE

An enduring theme in cosmology is the question of meaning. Do we matter? Does the universe know or care about us? What is it for? What's it all about? Physicists' emphasis on mechanisms and laws have given physics a reputation for being almost inhuman, perhaps pointless, certainly soulless. Atomic bombs and nuclear accidents have not helped its reputation. This alienation leads many nonscientists to recoil from embracing physics as what it is, a unifying system for describing nature, because we are supposed to be part of the universe, and none of us is inhuman, pointless, or soulless.

Wrapped up with this is discomfort with the idea of unyielding mathematical rules generally. Nobody wants to feel like a robot. Are we so unimportant that even our thoughts are dictated by the blind action of physics in our brains? There is a strong emotional attachment to the self and to the idea of free will.[2]

This was not always a problem. Ancient ideas about cosmology imposed or projected whimsical and humanizing ideas, involving constructions from turtles to cows, and many magical interventions. They included spiritual and mystical elements that in modern cosmology are usually firmly excluded. It went without saying that humans are in the middle of everything and that the universe literally revolves around us.

Physical insight, starting with thinkers at least as ancient as Aristarchus and later revived in the early renaissance by Copernicus, put the sun in the physical center of cosmic motion and consigned humans to an undistinguished orbiting satellite. This collision with ancient anthropocentric spiritual dogma led to the Vatican's prosecution of Galileo, but the compelling precision of mathematical physics gradually overcame the ancient prejudices; Kepler's laws of planetary motion and then Newton's laws of physics established a precise mathematical paradigm for explaining how and why things happen in the physical universe. Darwin ex-

---

[2]A stormy debate over the relationship between mind and brain continues now in the philosophical and neurobiological literature. I do not address these issues in the present discussion, except to assume that physics contains a complete description of physical phenomena—which is almost tautologically true.

tended the scientific approach by providing a rational explanation of natural history. We know now that the origin of species requires no divine intervention, nor does the origin of elements or cosmic structure. We are so comfortable with the separation of spiritual and physical elements in our cosmology today that the Vatican sponsors advanced scientific workshops on physical cosmology and has even endorsed Darwinian evolutionary theory.

Curiously, modern physical cosmology could nevertheless offer a roundabout way to restore humans to a central position in the cosmos. Cosmology allows the fascinating possibility of many parallel universes, each with its own set of physical laws. This possibility has inspired the "anthropic" hypothesis: "The universe" is not exactly made for us, but out of the many different parallel universes (which all "exist somewhere"), we cannot live in any but a habitable one, which therefore will seem customized to our needs. This certainly places us in an important position relative to our own universe: Our requirements have "selected" the physical laws for the entire universe. This idea cannot easily be contradicted, because the other, parallel universes cannot ever be observed, even in principle.

One might similarly argue that our requirements have selected for us, out of the vastness of space, a nice comfortable planet to live on, but of course this would not be the whole story: We have evolved to be perfectly suited to our situation, so naturally it seems very comfortable. On the other hand, we do live in conditions that are very atypical in the cosmos, and this is no accident—

life could not have evolved in the chemically brutal environment of interstellar space. It requires environmental nurturing.

The extreme alternative to the anthropic hypothesis is that this universe (the observable one we live in) is the only possible one, its properties completely dictated by mathematical necessity, a posture that takes the idea of Newtonian constraint to the limit. This view is supported by the undeniable existence of very tight, purely mathematical constraints on what is possible. The rules of physics contain perfect symmetries that are absolutely always true everywhere, even where they have no effect on our existence. We can, for example, measure transition energies of very distant atoms with high precision (e.g., using data such as Figure 24), and they are just the same as those of atoms at home. Possibilities are apparently highly constrained by very precise rules; the question is whether they are so constrained that there is only one possible set of physical laws.

(Note that even if our Universe is unique, our previous discussion about counting infinities shows that we are not "predetermined." Even in the most extreme case, only the global statistical properties of the universe and its physical laws are predetermined, not the detailed flow of microscopic information in the form of stars, DNA, neurons, and so on, which is forced by combinatorics into never-ending novel improvisations.)

Within the strict rules we know about, there is still room for many different universes with different laws, because there are many currently arbitrary tunable parameters of Standard model physics. For example, many particle masses could be different, as far as we know, without changing any other rules, in which case a lot of nuclear physics and chemistry would change; or there could be a small cosmological constant without affecting anything that we can think of except giving a small kick to the cosmic expansion. But these parameters might also be explained eventually through pure mathematics, a "theory of everything" such as Supersymmetric Superstrings—at least this is the hope of many who work on these theories. They expect that the particle masses can no more be adjusted than π or $\sqrt{2}$ can and that the cosmological constant really is exactly zero for reasons we can't yet articulate. If this ambitious unification program is successful, then we will have to reconcile ourselves to the fact that the universe does not exist to meet our needs, and that we have merely grown up to like it because that's how we got to be here, by adaptation—we are professionals at adapting to the situation, with experience of over three billion years.

There is also a middle ground: The laws may be almost completely determined by mathematical symmetry and yet still allow some variations among a limited set of alternatives in various universes, according to the values of a small number of randomly chosen symmetry-breaking parameters. The fact that such issues are unresolved means that we have to be very cautious about invoking arguments based on "beauty" or "naturalness." We have too limited a perspective on how nature does things to trust our intuition here.

In any of these "meta-cosmologies," it is no accident that we have emerged at a place and time interposed between the sterile symmetry of the largest and the earliest universe and the frenzied, fragile order of the microscopic world. We exist where there are the greatest physical and chemical opportunities for intelligence at this time. This is neither an accident nor evidence of our pivotal role in the universe; it is just a symptom of what we are and how we got here by accreting opportunities. It's not a surprise that we inhabit a congenial place in the universe, even if such places appear very few and far between.

Regardless of whether we are important to the global arrangement of things, we are likely to be very important to the development of the future potential of information. In our local region of space, we are at the vanguard of intelligent activity, and perhaps the local region extends as far as we can see.[3] The most important thing about us is what we create: our perceptions and technologies, our culture and societies, our art and our science. A billion years from now, if we have any cultural heirs, they will certainly not be biologically human, and they will probably not live on Earth, but they will have learned from us about their origin and ours—and they will still be telling some version of the story in this book.

---

[3]We do not know the distance to the nearest extraterrestrial intelligence; it may lie within our Galaxy, or it may be unobservably distant.

Suggestions for Further Reading and Browsing

The discoveries and insights explored in this book derive from the work of several thousand physicists, engineers, and astronomers. I have mentioned very few of them by name (and none living) as a way of avoiding a distracting secondary narrative about the historical development of cosmology; this modest primer has such a large scope that it would sink under the weight of sufficient scholarly attribution.[1]

---

[1]Before the 1960s, only a few dozen scientists worked in cosmology so it is easier to identify many key ideas and discoveries with individuals. Today the scientific community is much larger and more diverse, and attributions for the assembly of the Big Bang model and its observational support, if it were fair, would resemble the credits of several Hollywood movies, including not merely a list of the "authors" of the work but a listing of the scientific equivalent of producers and directors, actors and technical support, as well as financial backers such as universities, foundations, and governments. All of us of course owe a debt of gratitude to

Instead, I refer the interested reader to many excellent books about astrophysics and cosmology that add the historical dimension and provide entries into the technical literature.

The early developments were described by Arthur Eddington, *The Expanding Universe*, Cambridge, 1933; George Gamow, *The Creation of the Universe*, Viking, 1952; and Hermann Bondi, *Cosmology*, Cambridge, 1960. Steven Weinberg's *The First Three Minutes*, Basic Books, 1977, is a classic description of the microphysics of the early universe from a modern point of view; Joseph Silk's *A Short History of the Universe*, Freeman, 1994 is more comprehensive and up to date. There are several good recent biographical and anecdotal accounts of various aspects of current cosmological research, for example Timothy Ferris, *The Whole Shebang*, Simon and Schuster, 1997; Alan Dressler, *Voyage to the Great Attractor: Exploring Intergalactic Space*, Knopf, 1994; George Smoot and Keay Davidson, *Wrinkles in Time*, Morrow, 1993; Alan Guth, *The Inflationary Universe*, Addison-Wesley, 1997; and John Mather and John Boslough, *The Very First Light,* Basic Books, 1997.

———

(*continued*) the interested taxpayers whose support makes our work possible. Although practical benefits certainly result from cosmological studies (the most spectacular of which were Newton's laws of motion), they are motivated mainly by curiosity—it is probably not a coincidence that the global staffing level of the worldwide cosmological enterprise is comparable to that of science fiction entertainment. The cost of building a world class telescope such as the Keck I is about the same ($ 100M) as producing a blockbuster movie.

Steven Hawking's *A Brief History of Time*, Bantam 1988, discusses the nature of time and its boundaries in black holes and cosmology, and Roger Penrose's iconoclastic *The Emperor's New Mind*, Penguin 1991, includes an excellent description of quantum mechanics. Mitchell Begelman and Martin Rees' *Gravity's Fatal Attraction: Black Holes in the Universe*, Freeman, 1996, explains the astrophysical effects of black holes in quasars and elsewhere. Kip Thorne, *Black Holes and Time Warps: Einstein's Outrageous Legacy*, Norton 1994, includes detailed discussions of gravitational waves and of time travel facilitated by exotic forms of matter. For visionary thoughts on the potentialities of everything it's hard to beat Freeman Dyson, for example in *Infinite in All Directions*, Harper and Row, 1988; see also David Layzer's thoughts on structure formation in *Constructing the Universe*, Freeman, 1984.

John Barrow and Frank Tipler give a scholarly summary of the Anthropic Principle in their book of that title, Oxford, 1986, and Lee Smolin, *The Life of the Cosmos*, Oxford 1997, has presented a provocative specific model based on baby universes emerging from black holes. A superb recent synthesis and exploration of multiverses beyond our own is presented by Martin Rees in *Before the Beginning*, Addison-Wesley, 1997.

For those who wish to look at more pictures (or color ones), a good start for optical images is the Space Telescope Science Institute, http://www.stsci.edu/top.html. Sky maps of the background radiation can be found through the COBE home page, http://www.gsfc.nasa.gov/astro/cobe/cobe_home. html. Images and movies of simulated universes are posted by the N-body shop, http://www-hpcc.astro.washington.edu/.

For those who wish to go deeper (including a more serious commitment to mastering math and physics) the next step is to take freshman Astronomy 101 or to read one of many excellent general textbooks designed for these courses, for example, Hartmann and Impey, *Astronomy: the Cosmic Journey*, Wadsworth, 1994 or R.C. Bless, *Discovering the Cosmos*, University Science Books, 1996.

The most inspiring classic physics text is Feynman's *Lectures on Physics*, Addison-Wesley, 1989. Though these volumes are sometimes criticized for uncompromising depth and novelty, these attributes make them the best thing for anyone really excited by physics, despite their age. Another classic is Frank Shu, *The Physical Universe: An Introduction to Astronomy*, University Science Books, 1982, which gives good physical explanations of all the important discoveries in astrophysics, including those in cosmology.

To pursue this to the graduate level, classic texts of the cosmological renaissance are Dennis Sciama's still very readable *Modern Cosmology*, Cambridge, 1972, P.J.E. Peebles' notes on *Physical Cosmology*, Princeton, 1972, the second monumental volume of Zeldovich and Novikov, *Relativistic Astrophysics*, Chicago, 1983 (translation of 1974 Russian original), and Steven Weinberg, *Gravitation and Cosmology*, Wiley, 1972. More recent advanced texts on the subject are good places to get oriented in the scientific literature, for example Kolb and Turner, *The Early Universe*, Addison-Wesley, 1990, and P.J.E. Peebles, *Principles of Physical Cosmology*, Princeton, 1993. Good tutorials may be found also in Eric Linder,

First Principles of Cosmology, Addison–Wesley, 1997; Hawley and Holcomb, *Foundations of Modern Cosmology*, Oxford, 1998; Coles and Lucchin, *Cosmology: the Origin and Evolution of Cosmic Structure*, Wiley, 1995; and Michael Berry, *Principles of Cosmology and Gravitation*, Cambridge, 1976.

Original research in cosmology is reported not in books but in articles. A guide to the original research papers may be found in review articles, for example in *Annual Reviews of Astronomy and Astrophysics*, *The Review of Particle Physics*, published by Physical Review D, and *Reviews of Modern Physics*, and occasionally in the journals *Nature* and *Science*. A good database of the professional astrophysics literature is maintained by NASA at http://adswww.harvard.edu and a good access to physics literature (including online reviews) is maintained by the particle data group, pdg.ldl.gov. If you want to read about the latest raw results, straight from the authors and with nobody to help you sort out what is worthwhile, try the electronic preprint archive, http://xxx.lanl.gov, but remember—don't believe everything you read!

Index

Index

Index